Ecological Interface Design

Ecological Interface Design

Catherine M. Burns

John R. Hajdukiewicz

CRC Press
Taylor & Francis Group
Boca Raton London New York

CRC Press is an imprint of the
Taylor & Francis Group, an **informa** business

CRC Press
Taylor & Francis Group
6000 Broken Sound Parkway NW, Suite 300
Boca Raton, FL 33487-2742

First issued in paperback 2019

© 2004 by Taylor & Francis Group, LLC
CRC Press is an imprint of Taylor & Francis Group, an Informa business

No claim to original U.S. Government works

ISBN-13: 978-0-415-28374-8 (hbk)
ISBN-13: 978-0-367-39414-1 (pbk)

Library of Congress Cataloging-in-Publication Data

Burns, Catherine M.
    Ecological interface design / Catherine M. Burns and John R. Hajdukiewicz.
      p. cm.
    Includes bibliographical references and index.
    ISBN 0-415-28374-4 (alk. paper)
    1. Human engineering. 2. Human-machine systems. I. Hajdukiewicz, John R. II. Title.

TA166.B895 2004
620.8′2—dc22   2004045492

Library of Congress Card Number 2004045492

**Visit the Taylor & Francis Web site at
http://www.taylorandfrancis.com**

**and the CRC Press Web site at
http://www.crcpress.com**

DEDICATION

To Kim

TABLE OF CONTENTS

LIST OF FIGURES

LIST OF TABLES

FOREWORD

Anyone who has ever stepped foot inside a complex sociotechnical system — a power plant control room, a hospital operating room, an aviation cockpit — to observe workers performing even comparatively mundane tasks is struck by how flexible human behavior can, and must, be. Were it not for people's remarkable ability to adapt to situations that were not foreseen by designers, these systems would cease to function. Yet, most frameworks for design rest on a theoretical foundation that does not accommodate this reality because they are based on concepts — tasks, actions, procedures, strategies — that only deal with the foreseen. If you do not know the initiating event, it is very difficult — by definition — to conduct a task analysis or even a cognitive task analysis. This does not mean that these well-known concepts do not have a place in our toolbox — they do — but it does mean that they are insufficient.

The Ecological Interface Design (EID) framework was deliberately developed to help designers provide workers with information support that can help them play the role of adaptive problem solvers, and this book represents an important and badly needed contribution in moving EID from a research enterprise to (potentially) a practical reality. The demand for such a book can be illustrated by briefly reviewing the origins of the EID framework (Vicente 2001).

During the 1960s, a research program in the Electronic Department of the Risø National Laboratory in Roskilde, Denmark identified "design for adaptation" as a critical challenge in improving the safety of complex sociotechnical systems. During the next two decades, a number of conceptual tools were developed to tackle this practical problem, although comparatively little effort was devoted to rigorous experimental evaluation. Beginning in the 1990s, an empirical research program was initiated to test the value of those conceptual tools more rigorously and systematically. EID has now been tested in a variety of settings (e.g., process control, aviation, computer network management, software engineering, medicine, command and control, and information retrieval), and the results generated to date have been quite promising, although several crucial questions have yet to be addressed, let alone answered (Vicente 2002). This body of literature now comprises over 150 articles and continues to generate new insights. At the same time, the EID framework has received some attention from industry (e.g., Itoh et al. 1995) and there is a growing interest in "how to do" EID. Yet, the existing literature does not do a very satisfactory job at addressing this need (e.g., Rasmussen 1994; Vicente 1999), despite the fact that technology transfer to industry is seriously hampered unless more people become skilled in the application of the framework.

As I see it, this book is intended to fill this crucial gap. Burns and Hajdukiewicz have done a masterful job at moving EID from a craft that can be performed only by a few insiders using a great deal of "black magic" to a more easily understood and practiced design skill that can be performed by more people with less reliance on unspecified creativity.* Therefore, this book will be of interest not only to designers in industry who want to apply EID but also to academics at universities who are interested in teaching EID to their students.

Burns and Hajdukiewicz are highly qualified to take on this task because they have extensive experience in using EID, both in academe and in challenging industry-scale design projects. They have wisely chosen to adopt an engineering approach, focusing on presenting many different design examples from a diverse set of sectors. They also adopted a balanced perspective, showing that EID was not intended to — and in fact, cannot — be a replacement for other systems analysis and interface design techniques. While EID appears to offer some unique value added, there are many important interface design issues that the framework does not address, so it is important to use it in conjunction with other cognitive engineering techniques. This book provides convincing examples to illustrate how EID can, and should, be integrated with other methods.

Burns' and Hajdukiewicz's writing is clear and they present new and innovative techniques that address the mystical "how to" question rather than adopting the more traditional academic approach of searching for "the truth" (whatever that might mean). In doing so, I believe Burns and Hajdukiewicz have succeeded in significantly lowering the amount of effort, inside knowledge, and "black magic" that it takes to apply EID. I predict that this book will be a welcome and highly sought-after resource for corporations and at universities.

---

* In the interest of full disclosure, I should declare that Burns and Hajdukiewicz were both Ph.D. students of mine.

Having said that, it is important to temper our enthusiasm by not losing sight of the big picture. When a new set of ideas, such as EID, appears on the intellectual landscape, there is a tendency for advocates to jump on the bandwagon and adopt an uncritical and overly positive attitude that sometimes borders on religious fervor. Burns and Hajdukiewicz have smartly avoided this territory. Passionately pursuing ideas is important — indeed, necessary — but we should not forget that the most important goal is to continue to learn, not to "be right". This book was written in this spirit, and that is how it should be read, not as the final word or a definitive ideological statement, but as an important and much needed pragmatic advance for a problem-driven research program that continues to generate new insights for both researchers and designers.

Kim J. Vicente
Department of Mechanical & Industrial Engineering
University of Toronto

REFERENCES

Itoh, J., Sakuma, A., and Monta, K. (1995) "An ecological interface for supervisory control of BWR nuclear power plants," *Control Engineering Practice*, 3:231–239.

Rasmussen, J., Pejtersen, A. M., and Goodstein, L. P. (1994) *Cognitive Systems Engineering*, New York: Wiley.

Vicente, K. J. (1999) *Cognitive Work Analysis: Toward Safe, Productive, and Healthy Computer-Based Work*, Mahwah, NJ: Erlbaum.

Vicente, K. J. (2001) "Cognitive engineering research at Risø from 1962–1979," in E. Salas (Ed.), *Advances in Human Performance and Cognitive Engineering Research, Volume 1* (pp. 1–57). New York: Elsevier.

Vicente, K. J. (2002) "Ecological interface design: Progress and challenges," *Human Factors*, 44:62–78.

PREFACE

When we started this book, two things were bothering us. We were getting tired of hearing the fairly vague statements that "Interface design is an art" and that "Interface designers are fundamentally artists." While we don't completely disagree with these statements, we think that they lead to the misconceptions that you cannot teach interface design and you cannot learn interface design. This was not a very promising position from which to begin a book on the topic!

We asked ourselves, "But people teach art, and people learn art, so why not interface design?" While there will always be exceptionally talented artists who have a magic to their art that defies description, the world is full of people who have managed to become passably good artists. They learn techniques of composition, brushwork and the use of other mediums, concepts of color, lighting, and perspective. They learn the works of master artists and major genres of art, how these works were constructed, and what makes them good examples of their type. They can create a good representation of a scene or a different style of genre.

This is the approach that we have taken in this book. We cannot teach you to be a world-class interface designer (though some of you may become so), but we can teach you some of the techniques and expose you to the examples that should help you to become a passably good designer in this field. This book is constructed in this manner: The first four chapters present the techniques of composition and brushstrokes of ecological interface design by exposing the analytical methods behind the designs, the most common graphical forms, and how these are pulled together to make a complete design or picture. The last chapters are case studies, presenting larger works to expose you to relatively good examples of various techniques or results in the field. The final chapter attempts to situate ecological interface design with other common styles and approaches for interface design; this will allow you to better understand how to combine approaches to particular problems.

This is not a theoretical book or a description of current research in the field. A better reference for the state of current research is Vicente (2002). This book is not meant to provide a complete collection of EID works to date and by no means does so. There are many fine examples that are missing. From a research perspective, many of the examples here are possibly premature in terms of analysis, evaluation, or standardization. These were not the criteria used in selecting the examples. The examples were chosen because they showed a solution to a certain part of the EID problem. Some examples are stronger in the analysis phase, and some are stronger in the graphical design phase. Our motivation was to select examples that demonstrated something unique, that people could learn from and use as they developed their own designs. We have tried to capture — in the sense of art that we mentioned before — the idea of learning from different genres.

This book is written for our students, for industry practitioners, and for all others interested in learning this approach to interface design. The book has been tried out in draft forms with several classes and has been reviewed by industry practitioners who design advanced displays. The major decisions in the writing of the book have been informed by their opinions. It was our students and industry practitioners who told us that they preferred more examples over highly researched examples, and that both analytical examples and graphical design examples were important for them to see. They wanted case studies, and summaries, and checklists, and minimal acronyms. We've tried.

C.M.B. and J.R.H.

ACKNOWLEDGMENTS

There are many people that we must thank for their contributions to this book. First, two classes of SYDE 542 (2002 and 2003) at the University of Waterloo and one class at the University of Toronto, taught by Greg Jamieson, kindly used incomplete drafts of this book as their text. We appreciate their patience and were overwhelmed by their generous comments and their interest in the future of the book. This book is really for them and we hope the comments will keep coming in case we get the chance to revise it. Notably, comments came from Gabe Chan, Analene Go, Munira Jessa, Lillian Mah, Teresa Mak, Samer Sawaya, Galang Vuong, Kristina Watt, and Shelley Zhou. We thank Greg Jamieson of the University of Toronto as well for his many comments on the book and his perspective on using it as a teacher.

From the Advanced Interface Design Lab at the University of Waterloo, we thank Ed Barsalou, Angela Garabet, and Laura Thompson for their comments.

We are grateful to Kim Vicente of the University of Toronto for writing the Foreword, his support over the years, and his comments on this book. It was Kim who started us in this area, and we are continually thankful for his leadership.

Numerous industry practitioners have reviewed whole or partial drafts of the book, providing an industry perspective. In particular, we thank Nick Dinadis, Jamie Errington, Lisa Garrison, Laura Gosbee, Dal Vernon Reising, and Bill Rogers for their extensive comments. A special acknowledgment goes to Lisa Garrison, who coined the term "visual thesaurus," which has led to one of the more interesting sections of this book.

Finally, we greatly thank our families, friends, and Jack for their continued support in our adventures.

C.M.B. and J.R.H.

LIST OF ACRONYMS

AH   Abstraction Hierarchy
FP   Functional Purpose
AF   Abstract Function
GF   Generalized Function
PFn   Physical Function
PFo   Physical Form
EID   Ecological Interface Design
PDA   Personal Digital Assistant
UCD   User Centered Design
WDA   Work Domain Analysis

In the interest of readability, and at the suggestion of our students, we have tried wherever possible to reduce the use of acronyms.

1

INTRODUCTION
AND OVERVIEW

Interface design is an evolving discipline. Many of the innovations that are possible in interface design today are a result of improvements in the technology of displays, graphics cards, and graphical software. The cycle of how we apply the technology, followed by new capabilities, followed again by a new application, sets up a continuous march of progress into the future. How to use these technological enablers to our advantage is an exciting question. As technology continues to evolve and remove constraints on what can be designed, computer interfaces should defy the imagination in terms of the creative ways in which they can support human work. It is against this backdrop of ever-evolving and improving technology that we present the ideas of Ecological Interface Design (EID).

EID is, at the time of writing this book, a relatively new approach to designing computer interfaces. EID is focused on the design of interfaces for large and complex systems, such as power plants and medical equipment. We will stretch the discussion of EID to social systems and to small-screen PDA-style displays, but there are relatively few applications of EID in these areas. The term "EID" and its formalized approach originated in a systematic form with the work of Vicente and Rasmussen (1989). In actuality, though, there were "ecological" designs before this time, and there are ecologically sound designs that continue to emerge from other design perspectives.

By an "ecological" or "ecologically sound" design we are referring to an interface that has been designed to reflect the constraints of the work environment in a way that is perceptually available to the people who use it. Simply put, the users are able to take effective action with the interface, understanding how those actions will move them towards their

1

objectives. When an ecological interface is implemented well, even complex relationships and data are easily visualized and used naturally. This transparency of use, where users feel as if they are working directly with the object and not with the interface, is the "holy grail" of interface design. If you are exploring EID after learning about other interface design approaches, we hope that you will see EID as a way to supplement your current approach with some new ideas on determining design requirements and some new ideas on displaying complex information relationships.

While there are many ways of achieving an ecological display, this book will focus on developing interfaces from what is called a Work Domain Analysis (WDA; Vicente 1999). This is the approach that spawned interest in EID and, in our opinion, is the most formalized and systematic approach available at this time to help you create an ecological display. You can learn more about the theoretical motivations and the method of WDA in a book by Vicente (1999) and the state of EID research (Vicente 2002). The objective of this book is to review WDA and concentrate on how it can be used in designing an interface. We have included many examples, including several case studies showing the analysis and subsequent implementation of WDA. Where possible, we have included a brief description of evaluation results. This book, though, is focused on teaching the method of WDA and how to apply it in an EID; it is not meant to be a review of empirical work in EID. So in all cases, we encourage readers with a stronger interest in the evaluation of the interfaces to pursue the original published papers for more details.

SCOPE OF THIS BOOK

This is a book on Ecological Interface Design and is focused on applications. It is intended to show how a WDA can be performed in different work domains and, then, how the results of that WDA can be effectively applied in design. This book is intended for people interested in developing ecological interfaces in industry, or for graduate or senior undergraduate students taking a special topics course in EID. Our motivation in this book is to bridge the gap between WDA and design, through the discussion of several examples. It is not our intention to concentrate on the theoretical underpinnings of EID; that discussion is available elsewhere (e.g., Vicente 1999; Rasmussen et al. 1994). This book is also not intended to be a review of empirical work by researchers in EID. This, too, is available elsewhere (Vicente 2002). Where available we have provided brief summaries of major results, but readers interested in this direction should seek out the original papers. As well, the examples in this book have been chosen because they demonstrate ways of handling analytical or design-related challenges that are useful for teaching the design of ecological interfaces. While we have included as many examples as possible, it has not been our intention to catalog the extent of EID work to date.

TYPES OF DESIGN PROBLEMS

We are surrounded by different interfaces, from computer websites to automobile interfaces, interfaces at work, and even phone or PDA interfaces. While all of these interfaces present design problems, they present design problems of different kinds. The kind of interface that you want to use to drive your car is very different from the kind of interface that your mechanic uses to diagnose an engine problem with your car. Similarly, the interface on your computer has been designed for different purposes than the interface that an anesthesiologist might use to monitor you during surgery.

Some of the key differences between these design problems are:

■ The degree to which the user must engage in diagnostic problem solving
■ The complexity of the work domain
■ The expected skill level of the user
■ The degree to which the work domain is safety critical or time dependent

While you certainly want your car mechanic (or anesthesiologist!) to quickly diagnose and correct problems, when you drive your car and are dealing with the hazards of roads and traffic it makes sense to keep the car interface fairly simple and restrict indications to those that are most critical. You usually have more time to solve problems, and in most cases you are interested in getting where you need to go and intend to leave fixing the problems of your car to someone else. You are not an expert in the domain, and an interface designed for an expert would seem overwhelming in its detail and its complexity. In contrast, anesthesiologists are a highly trained group of individuals with high levels of expertise. While novice users often prefer an interface that is simple to use, experts in jobs requiring diagnosis and problem solving require access to, and usually prefer, interfaces with as much information as possible (e.g., Tufte 1997; Burns 2000).

Whereas a quick survey or focus group session can elicit the preferred layout of controls for a car radio, understanding the information requirements for the mechanic and the anesthesiologist requires extensive study by the designer. Though EID may certainly give insights to other kinds of design issues, EID has been developed to aid designers who must design interfaces for these complex, safety-critical systems where expert users must engage in diagnostic problem solving. These are the kinds of examples that you will see discussed in this book. We will look at examples from transportation systems, process control systems, telecommunications systems, medical systems, and social systems. Finally, we will look at how EID can be used with other design methods.

THE ECOLOGICAL PART OF EID

At first glance, "ecological" may seem an odd term to pair with "interface design," bringing to mind thoughts of biology and ecosystems. The actual origin of ecological in EID refers to a small field of psychology called "Ecological Psychology" (Gibson 1979/1986). Ecological psychology studies human–environment interrelationships and human perception in rich environments. This contrasts with experimental psychology, which studies human behavior in laboratory environments and, to a certain extent, with perceptual and cognitive psychology, which is less focused on human–environment interactions. Ecological psychology advocates that human behavior is often constrained by the environment that we work in. Because we want to design for people working in complex environments, and not laboratories, and we know that their environment presents them with certain constraints to work within, ecological psychology provides a useful psychological foundation for interface design. Three of the key tenets of ecological psychology that we use in interface design are:

- People's actions are constrained by their environment or work domain, so the work domain must be understood before starting a design.
- It is possible to design interfaces (or "mediated environments") that provide information that people can pick up and use.
- There are visual ways of displaying information that can reduce the need for memory or mental calculation.

These three tenets underlie the EID approach. First, we maintain that people must make decisions that are constrained by their work domain; second, we maintain that we can systematically analyze these work domains to determine these constraints; and third, there are design techniques and visualizations that can show these constraints in a way that reduces the need for mental calculation or memory.

OUTLINE

This book can be divided into two sections. Chapters 2, 3, and 4 introduce the WDA and discuss how to use it to develop a display. Chapters 5 through 9 discuss several case studies of EID, looking in detail at how the analysis can be performed and implemented in designs; they also give the results of these EID projects. Chapter 10 situates EID in the context of other design methods. While the book can be read straight through, it is intended that the example chapters stand alone as a domain-based reference on the application of EID.

MOTIVATION FOR EID

For people who have seen the success of User-Centered Design (UCD), the question must come up, "Why even use EID at all?" The UCD perspective has certainly advanced the design of user interfaces and led to great improvements. The general success of the "windows"-based interfaces used in most current computing environments is a good example. Before UCD, interface design was often the job of computer programmers, with interfaces usually being added on as the finishing touch to a program. UCD was quite a revolutionary change in perspective, most importantly adding three key ideas:

1. **That interface design is a field on its own because it bridges between humans and the program/environment**
2. **That an understanding of human perception, cognition, and behavior is critical to designing interfaces**
3. **That much can be learned by getting feedback from the actual users of the interface, at the early design stages, and then through testing at various points in the design**

The UCD perspective moved interface design from the task of computer programmers to either programmers with extra training in psychology or specialist user interface designers. To a large extent, the results have been very successful.

There are reasons, though, for being interested in learning the EID approach. EID is useful for certain kinds of problems and in certain kinds of situations. The key times to add EID to your design process are:

- When asking users doesn't work
- When we want users to become experts
- When we want to handle the unexpected

We discuss these situations in the next sections.

When Asking Users Doesn't Work

The problem with UCD results when we deal with more complex systems. In systems as complex as a power or petrochemical plant, users, even very experienced operators, do not have a complete understanding of how the plant works. You can't simply ask them, "What do you need on the display?" and expect the answer to be complete and consistent across users. If you ask five different users, you will get five different answers.

The other reason that UCD doesn't always work in these situations is that there are constraints underlying how the domain works. By constraints, we mean a set of relationships that must hold for the system to be working correctly. If all the constraints are satisfied, the system is working. If one or more constraints are not met, some aspect of the system is broken. Sound easy? Not really. Users are not always aware of the constraints that affect the system they work with, and discovering these constraints can take some extra effort. It can be well worth it, however, because understanding these constraints, and making them visible, can help your users to use your interface far more effectively. They might even learn more about the system than they knew before, just from using your interface.

To give an analogy, let's consider trying to book a meeting between five people in four different time zones: ET, CT, MT, and PT. As a first approach, you might send out a memo or email to all five people asking them when they would like to meet. More than likely you will get five different answers: One person wants to meet Monday at 9:00 AM ET, the other Tuesday at 10:30 AM MT, and so on. For the person at the end compiling these answers, the job may seem impossible, and very likely a solution won't be reached this way. Applying UCD to a complex system is similar to this; users will give you different answers and you won't uncover the true constraints of the system.

In contrast, another way to handle the problem is to start with the constraints first, and then seek user input. In our meeting problem there are most likely some basic constraints. Most people will want to meet during typical business hours, say 9:00 AM to 5:00 PM. The time zones add another kind of constraint, reducing the possible meeting hours to 12:00 PM to 5:00 PM (ET), or 9:00 AM to 2:00 PM (PT). By understanding the constraints of the problem, we can reduce the scope of the solutions to those that are truly feasible. This improves the chances of getting a working solution. At this point, users can be asked for input, or if you wish, you can even pursue another layer of constraints (e.g., holidays or other commitments) and reduce the scope of the problem further. If you have tried to book a meeting with multiple people, you may have naturally adopted this constraint-based style. EID is very similar; it examines the constraints of the work domain before getting user input. This isn't to say UCD is wrong or ineffective, but simply that there are different kinds of interface problems that require a different approach. Once the constraints are known, the principles of UCD are invaluable for completing your ecological design effectively.

So if we were to go back to our list of UCD contributions, there would be one more key contribution that EID can add:

4. **In complex systems, it is useful to understand how the system works, before beginning your design.**

When We Want Users to Become Experts

Unique to complex systems is the fact that users don't always understand all the relationships behind the systems they work with. These systems are often large, highly interconnected, and with unpredictable dynamics. Learning to operate these systems can take years of training as skills and knowledge develop. If we, as designers, can do some of the work of understanding these systems and encapsulate that knowledge in the user interface, we can actually reduce that learning curve, or help the users to reach levels of expertise that were not possible before.

In order to do this, it is necessary for designers to move outside the user community to get input for their designs. It can mean talking to the original designers of the system, consulting drawings, textbooks, or engineers in the field. When we do this, we provide our users with a greater level of support than if we just surveyed users, as in a UCD approach. We give our users interfaces that allow them to learn more about the systems they work with and that can change the way they think about their work. That's absolutely required when designing for complex systems.

The other reason we want to do this gets back to supporting different kinds of user behavior: skill-, rule- and knowledge-based (KB) behavior (Rasmussen 1983). To support KB reasoning we need an interface that is rich in information and structured to show the relationships between that information.

When We Want to Handle the Unexpected

Unanticipated events are exactly that, unexpected. We can't design for unanticipated events by recalling events we have experienced before. We can't design for unanticipated events by trying to guess at what could happen. In fact, if we design to fit certain scenarios we run the risk of developing a design that is not flexible enough to support situations that are different from that scenario. As an example, consider the task of trying to keep squirrels from nesting in an old house. We can anticipate that the squirrels may crawl in through vents in the attic and block those openings. A neighbor may tell you that squirrels nested in their unused chimney. To go around and block each possible entry spot is taking a scenario-based approach to the design of the house and the prevention of the squirrels. As most homeowners know, the squirrels will eventually get in the house in a way they never expected. That's an unanticipated event.

One way to design an interface that can handle these situations is to design it based on constraints. When we design based on constraints, we can handle unanticipated events because, regardless of the event, the constraint is broken. If we define the squirrel problem as getting past the exterior of the house, we have defined it in a constraint-based way. With this kind of definition, it doesn't matter whether the squirrels enter through the vents, or crawl under the eaves, or chew through the bricks. The problem has been defined in a new way that covers these situations. When we do this with interfaces, we show the user the constraints of the system when it is working properly (i.e., the exterior integrity of the house). Any break in the constraints indicates that an event (i.e., a squirrel attack) is taking place. Furthermore, now we have also shown our users how to act. Restore the constraint and the system is fixed.

HISTORY

EID has evolved from different theoretical branches. While it draws on ecological psychology (e.g., Gibson 1979/1986), the true roots of EID came from a control engineering perspective that developed in Denmark in the 1960s. Vicente (2001) provides an extensive historical perspective of research at Risø. This summary has been constructed from his work. The Risø National Laboratory was formed in 1956 with the aim of conducting research to develop nuclear power plants in Denmark. In the early 1960s, the head of the Electronics Department, Jens Rasmussen, was involved in issues surrounding hardware reliability for nuclear power plants. These studies of hardware reliability shifted to an interest in human reliability with the inevitable realization that, in a complex system, the human operator is an important part of overall system reliability. From this perspective on complex systems, a program of study evolved that looked at the interaction between human operators, equipment, and automation in complex systems. Cognitive engineering as a discipline had been born. Later, in 1988, Norman would popularize the term "cognitive engineering" in his wide-selling book, *The Design of Everyday Things* (Norman 1988).

The Risø researchers analyzed industrial accidents and came to the conclusion that most accidents began with "nonroutine" operations. Human error and poor control choices were often cited as a cause or an extenuating factor in these analyses. From these conclusions, it became apparent that what was badly needed was an approach to design that could encompass abnormal situations and reduce human error. These requirements are still relevant today (e.g., Perrow 1990).

The Risø researchers began a program of studying operator performance. From this they developed two key ideas. First, that operators demonstrate three distinctly different kinds of behavior when interacting with complex systems. Some of their behavior is "skill-based," meaning that it depends on automatic responses and neuromuscular control. This is the kind of behavior that is needed to physically make a control action. Some behavior is "rule-based," or can be described by "if-then" inferences about plant state and appropriate actions to take. Rule-based behavior is dominant in routine or proceduralized situations. Finally, some behavior is "knowledge-based," or requires operators to reason about how and why the plant works. This is the behavior that is required to solve problems and to control the plant in unanticipated situations. These three kinds of behavior were combined in a taxonomy of human behavior that came to be known at the "S-R-K Taxonomy" for "Skills-Rule-Knowledge." The argument of this taxonomy is that all three kinds of behavior need to be supported when designing an interface for operators of complex systems.

Clearly, knowledge-based reasoning is critical to correct operation in accident situations, especially unanticipated situations. For this reason, the second direction that the Risø researchers investigated was human problem solving. They conducted field studies of electronic troubleshooters, mapping out the chain of reasoning that these troubleshooters followed as they solved problems (Rasmussen and Jensen 1974). From these studies, they realized that people reasoned about problems using a structure — basically asking "How" (How does this work?) and "Why" (Why is this here?) questions as they worked through the problem. This collection of "how/why" (in other words, "means-ends") relationships could be brought together to form a hierarchy of relationships (Rasmussen 1985). Because each level of this hierarchy describes the system at a different level of abstraction, this hierarchy was called an Abstraction Hierarchy.

EID has adopted the Abstraction Hierarchy as a fundamental way to analyze the environment, the work domain. By gathering these relationships in the form of information requirements, designers can create interfaces that will aid problem solving and retain their robustness in unanticipated situations. The information they are collecting matches directly with the kind of information that has been found to be useful in problem solving. This is a key idea behind EID and what makes ecological displays different from other kinds of displays. Chapter 2 will expand on how to do a Work Domain Analysis, how to create an Abstraction Hierarchy, and how to collect Abstraction Hierarchy information.

Armed with a conceptual understanding of operator performance, the SRK Taxonomy, and an analytical tool for analyzing work domains, the Risø work became noticed by the U.S. nuclear power industry in the wake of the Three Mile Island (TMI) disaster in 1979. One of the factors behind the TMI accident was that the operators could not correctly diagnose the problem or understand the state of the plant given the information that was available on their control boards (Perrow 1990). A valve that the operators thought they had closed had, in reality, stuck open. While the operators could see the control signal they had sent to the valve — that they had asked the valve to close — information on the true state of the valve, or flow through the valve, was not available and would have shown that the valve was still open. It was hours before the true state of the plant could be diagnosed. A new approach to determining information requirements for design was clearly needed.

Throughout the 1980s computer technology became cheaper, faster, and quite simply, better. Notably, monitors and graphics cards improved dramatically, moving from the early monochrome green or amber displays to low-resolution 8-color displays, and eventually to high-resolution displays capable of millions of colors. Not only did the displays get better, but graphics adapters were able to drive them faster and show far more complex graphics than ever before. Computer interface design was finally able to develop technologically as a field.

Against this backdrop, the idea of EID emerged. First named by Rasmussen and Vicente (1989), Vicente and Rasmussen (1990, 1992) combined the analytical tool of the Abstraction Hierarchy with the insights of the SRK Taxonomy to develop an approach to interface design for complex systems where unanticipated situations were a reality. Vicente developed a small process control microworld he called DURESS (and later DURESS II when it was expanded to include control). This environment allowed the new ideas of EID to be tested and explored. The environment was small enough that experiments could be run in a reasonable length of time but could be complex enough that performance differences and different strategies could be elicited. Two interfaces were developed for this system: an EID interface and a non-EID interface. Figure 1.1 shows the EID interface for DURESS.

Vicente's research program continued at the University of Toronto throughout the 1990s and continues today. The EID research program, using DURESS II, has improved the understanding of how ecological interfaces affect operator performance. Consistently, they have found that operators diagnose problems more quickly and accurately with ecological interfaces, which is consistent with the basis by which ecological interfaces are designed. They have found that operators adopt more flexible control strategies (Pawlak and Vicente 1996) and, to a certain extent, think about the plant at a higher level of abstraction with ecological displays (Janzen and Vicente 1998; Hunter et al. 1996). In short, they have confirmed that it is possible to improve operator performance and develop more reliable human–machine systems through this approach.

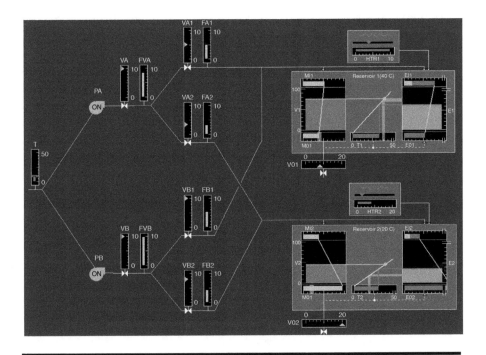

Figure 1.1   EID interface for DURESS. (Reprinted from Pawlak, W.S. and Vincente, K.J. (1996) Inducing effective operator control through ecological interface design. *International Journal of Human-Computer Studies,* **44: 653-688. ©1996 with permission from Elsevier.)**

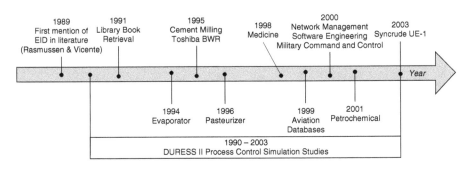

Figure 1.2   Historical development of EID.

Interest in EID has been expanding since the 1990s, and the approach has been applied to many more domains than process control or nuclear power plant control (Figure 1.2). To a large extent, EID has gained favor as a design approach for interfaces for large and complex systems. EID has been applied to medical systems (Sharp and Helmicki 1998; Hajdukiewicz et al. 2001), network management (Kuo and Burns 2000), aviation (Dinadis and Vicente 1999), and petrochemical systems (Jamieson and Vincente 1998; Jamieson 2002). These systems all share similarities in complexities, uncertain dynamics, and the need to provide support for operators in unanticipated situations. In this book we will review several of these applications of EID, explaining the analysis for that type of system, and showing how the analysis has been translated into an interface. There have been some applications of EID to less safety-critical systems as well, and some benefits can be realized in these kinds of work domains.

2

WORK DOMAIN ANALYSIS

In order to incorporate domain information into our design, we need a systematic way of searching for this information: asking the right kinds of questions, and storing these relationships in a way that we can use later. We need to know that we have covered different kinds of constraints and have achieved the adequate breadth of coverage. Rasmussen (1985) proposed a framework for understanding work domain constraints that he termed the "Abstraction Hierarchy." The Abstraction Hierarchy is the key tool in performing a Work Domain Analysis (WDA). A WDA consists of an Abstraction Hierarchy done at various levels of detail. A Part-Whole Hierarchy determines these levels of detail. In this chapter, we work through the following ideas, defining a system boundary around a system of interest, the basics of an Abstraction Hierarchy, and using an Abstraction Hierarchy along with a Part-Whole Hierarchy to complete a full work domain model. At the end of the chapter we discuss models with social constraints, managing large models, and some ways of double-checking and testing your work domain model for completeness.

DEFINING THE SYSTEM OF INTEREST

In WDA, the system of interest is defined differently than in other human-centered approaches. Whereas other approaches start from the user, WDA begins by examining the environment. This environment is the system that will be controlled by the user. Determining this system boundary is the first critical element of any analysis. Where you draw the system boundary will influence the kinds of displays you develop and the types of problems and tasks that users can perform with your displays.

The key aspect of defining your system boundary is to include anything that your users might want to control or have information on within the boundary and to leave anything you want to design or redesign outside of the boundary. This means that the interface is certainly left out of the analysis, but also usually databases, signal acquisition systems, networking equipment, bar code readers, input devices, and output devices. These items are usually peripheral to the main system that the user wants to control. We want these items to be as transparent as possible, so that the user sees the work environment and not the intermediary technology. If the user (such as a network manager or a database administrator) must monitor these devices, intervene, and troubleshoot them, then these items belong in the work domain. But if the user would not typically control these devices, or would contact someone else to troubleshoot the devices, we would not include these in the work domain. For example, a doctor may use a database with patient records, but the primary work domain is the patient. We do not intend for the doctor to be a database administrator or to be troubleshooting the computer network. This is what we mean by determining the main design domain and separating it from peripheral elements with a system boundary.

The second stage in defining your work domain model is to determine the scale of the problem you need to solve. There are key questions at this point:

1. Is your user focused on one piece of equipment?
2. Is the key problem integrating across several pieces of equipment?
3. Does your user monitor and supervise other people?

In the first case, the system boundary would be drawn around a single piece of equipment. The analysis would be tight and focused. In the second case, the analysis must include all the related pieces of equipment. In the third situation, the analysis would be broader. Multiple pieces of equipment would be included, as well as possibly supervised human components. The analysis would be directed towards a larger definition of work.

Determining these system boundaries does affect your analysis and, ultimately, your design. A boundary that is too tight will leave out important aspects of the work domain. Conversely a boundary that is too broad will waste time analyzing areas that are not of interest. When in doubt, though, you can start with a broad analysis and then concentrate the analysis on a specific problem area.

───────────────── ▼▲▼ ─────────────────

SYSTEM BOUNDARY

What to leave out (usually)
Databases, data storage devices
Signal acquisition devices, input devices, output devices
The interface, interface software, interface elements
Anything that you are designing or redesigning

What to leave in (usually)
Things the user can control, wholly or partly
Things that interact with the user's work domain
Things the user must monitor or supervise

WHAT IS AN ABSTRACTION HIERARCHY?

Generally, a hierarchy is a tree-like structure with multiple levels. The levels are distinctly different from each other and are ordered along some dimension. An Abstraction Hierarchy has the same components:

1. Levels
2. Tree-like structure
3. Levels are different and ordered along a dimension

There are lots of different kinds of hierarchies. One example is the organizational structure at your workplace or university. Using an organizational structure, you might have a hierarchy as shown in Figure 2.1.

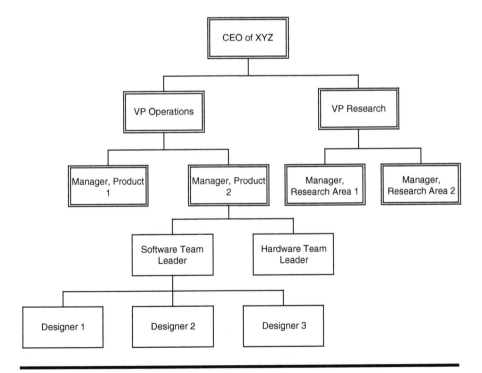

Figure 2.1   A corporate hierarchy.

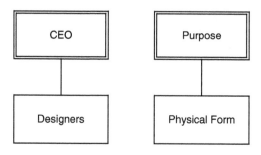

Figure 2.2   Anchors on a hierarchy.

We are all familiar with hierarchies like this. The people at each level perform different roles, but roles that are relatively similar to those in their same level. Even the names of the positions will tend to be similar within a level. The links up and down the hierarchy define responsibility and reporting. The higher levels are responsible for the people below them, and everybody reports up one level.

An Abstraction Hierarchy is very similar, except that it describes various elements of the work domain, and those elements are ordered along the dimension of abstraction. By abstraction, we mean how closely that element describes the physical nature of the work domain. The closer the connection, the lower the level of abstraction. If an element describes something closer to the purpose of the work domain, we say it is part of a description at a high level of abstraction. Just as the CEO and designers are the two anchors in the organizational hierarchy in Figure 2.1, purpose and physical form are the anchors of the Abstraction Hierarchy (Figure 2.2).

The other levels of the Abstraction Hierarchy, like the other levels of management, describe intermediary levels between purpose and physical form. In most cases, we use five different levels (Figure 2.3) to describe a work domain.

Like with the organizational chart, each level is different, and the connection between the levels is defined and consistent. In this case, the levels are connected by what are called "means-end" links. Means-end links are "how/why" links, with each level below explaining how the level above is achieved.

The Abstraction Hierarchy is a derivative of an engineering technique called Functional Decomposition (Ulrich and Eppinger 2000). Functional Decomposition is used in the design of new products to ensure that the most effective product has been designed. In Functional Decomposition, engineers start by asking, "What problem do we want to solve?" (Functional

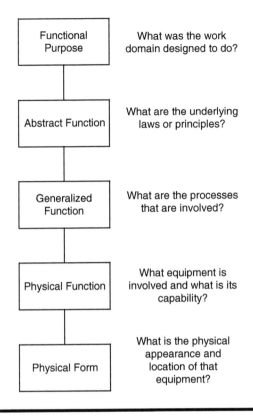

Figure 2.3   The five levels of the Abstraction Hierarchy.

Purpose) and then progressively ask "How?" as they move through the various levels. At the first "how" (Abstract Function), they look for the means to accomplish the purpose, usually through mass and energy transformations. At the next "how" (Generalized Function), they look for processes to accomplish those transformations. They look for equipment to accomplish those processes (Physical Function), and then they build the design (Physical Form). The Abstraction Hierarchy is identical to Functional Decomposition except that Functional Decomposition occurs before something is built; interface designers usually design after something has been built. This means we are constrained to the "as-built" system. When we build a work domain model, we are, in effect, re-creating some of the steps behind the original design of the work domain. When we understand why the work domain was built and how it works, we have gained a lot of useful information to use in our interfaces.

──────────────▼▲▼──────────────
FOCUS ON TERMS

Functional Decomposition — an engineering technique used to generate design alternatives, pre-design.
Abstraction Hierarchy — a post-design analysis that describes how a system works.
Part-Whole Hierarchy — an analysis that breaks apart the system into subsystems and components.
Work Domain Analysis — a complete analysis, Abstraction Hierarchy + Part-Whole Hierarchy combined.
Work Domain Model — the model that results from a work domain analysis.
──────────────▲▼▲──────────────

LEVEL BY LEVEL, WORKING THROUGH AN ABSTRACTION HIERARCHY

Abstraction hierarchies vary with the kind of work domain being analyzed, but there are still many similarities between different abstraction hierarchies. In general, the levels, or rather the type of information, is consistent across different hierarchies. In this section, we discuss the Abstraction Hierarchy level by level.

Functional Purpose

At Functional Purpose, we want to determine what the work domain was designed to do. This is sometimes called the "designed-for purpose" of the work domain. If we were to do a WDA of a car, the question we would be asking is, "What was the car designed to do?" The kind of answer we would be looking for might be, "Transport people from A to B quickly" and "Transport people from A to B safely" (Figure 2.4). We are looking for a generic description that holds over all the various tasks of a car. In contrast, the tasks of a car may be "Drive to work," "Drive to the store," and so on.

Besides being generic, a Functional Purpose description needs to include some evaluative criteria that can help the user determine if the work domain is functioning correctly. In the previous description we already identified several criteria:

1. The car must succeed at leaving point A and arriving at point B.
2. The transportation needs to be quick.
3. The transportation needs to be safe.
4. People are what is being transported.

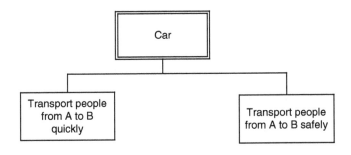

Figure 2.4   Functional Purpose of a car.

Therefore, if a car does not succeed in getting to point B, travels very slowly, endangers its occupants, or cannot transport people, then it is a failure at being a car! In other words, something has gone wrong with the car. We could define these criteria even more specifically to say the car must transport up to four people from A to B at up to 120 km/hr with no physical damage to the people. But for the purposes of these examples, the more general criteria are adequate.

FUNCTIONAL PURPOSE: QUESTIONS TO ASK

1. What was the work domain designed to do?
2. How do I know if it is working correctly?
3. What is good performance as opposed to bad performance? Do my purposes express these criteria?
4. Have I found at least two purposes?
5. Are my purposes generic? Do they hold across all possible tasks?

The other thing you will notice with the Functional Purpose is that we defined two distinct yet potentially conflicting purposes. If the car, for example, traveled very fast but injured its passengers, then it would not be a successful car. Defining at least two purposes is a useful thing to do in your analysis. First, most work domains have been designed for more than one purpose, and second, two purposes can be used to show the trade-offs and constraints between elements of the work domain.

Distinguishing Purposes from Tasks

Cognitive task analysis, hierarchical task analysis, and physical task analysis are human factors methods that define and analyze tasks. WDA is different from these approaches in that it searches for information on how the environment works, regardless of task. These approaches are complementary, if used properly. If you include tasks in your work domain model, though, you won't get the benefits of the different kind of information that this analysis extracts. So, what is the difference between a purpose and a task?

A task is an action or a series of actions that people do. A purpose is an attribute of an object or system. Tasks change throughout the day and throughout the year. Purposes stay constant. We learn about tasks through observation. Purposes are more often obtained by questioning or defined by reflective thought.

An analogy can be used to illustrate this distinction. There are many tasks a person needs to complete when building a house. But a house itself has a certain purpose: to provide shelter. This purpose is an attribute of a house. A house that does not provide shelter is no longer a house. So, in this way, a model of how someone could build a house would be a task model. A model of what is needed to have what we consider to be a house — some sort of foundation, support structure, and top covering or roofing — would be a work domain model. The work domain model would tell you that the purpose of a house was to provide shelter from the environment. The task model might tell you the steps and order that you need to do things in; the work domain model would tell you what weight and type of roof you could support, or how strong your beams needed to be. Obviously, both kinds of information are needed to actually build a house, to make sure that what you have built doesn't fall down, and to ensure that it functions correctly.

In multitask domains, it can be particularly difficult to determine the purpose of the domain. One technique is to list the various tasks that occur, and then ask what purposes are common to those tasks. If we worked with the house example above, we could construct a grid as in Table 2.1, with tasks in rows and purposes in columns.

The objective of the table is to establish a means of connecting diverse tasks. The way the tasks connect is through purposes. As the defined purposes are shown across multiple tasks, this begins to confirm that they are overriding attributes of the environment.

Table 2.1 Converting Tasks to Purposes

Task	Shelter from Rain	Maintain Livable Temperature
Pour foundation	x	
Set frame	x	x
Attach roofing	x	x
Connect gas lines		x

Abstract Function

Abstract Function is a description of the causal relationships underlying the work domain: the laws that cannot be broken and the priorities that must be achieved. In most cases, we are describing the laws of physics at this level. For example, conservation of mass tells us that whatever mass enters a system, it must either be stored in that system or must exit the system. Mass cannot be created or lost by a system.

EXAMPLES OF CASUAL RELATIONSHIPS THAT SHOULD BE DESCRIBED AT ABSTRACT FUNCTION

Conservation of mass
Conservation of unchangeable resources
Conservation of energy (1st Law of Thermodynamics)
Net increase of entropy (2nd Law of Thermodynamics)
$F=ma$
Conservation of momentum
Atomic balances of chemical equations

The other things we describe at the Abstract Function level are the flows of things that are conserved. In other words, we would describe how mass moves through a system, always ultimately being conserved. Similarly, we would look at how energy moves through a system. One way to determine what should be represented at Abstract Function is to ask, "What things flow through the system?"

THINGS THAT FLOW

Mass
Energy
Information
Money
Forces
Unchangeable resources

Your Abstract Function level must represent the fundamental nature of your work domain. For example, a power plant is ultimately about producing energy as electricity, so the Abstract Function level must describe the energy flows and relationships of this system. Similarly, a

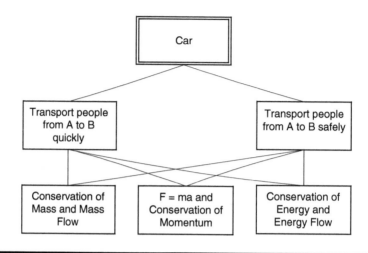

Figure 2.5   Abstract Function level, car example.

stove is also an energy system. In contrast, a car is a transportation system; it is fundamentally about moving mass. Mass movement and force generation are key relationships to model at this level, shown in Figure 2.5.

After modeling the fundamental causal relationships, secondary relationships become apparent. For example, power plants use mass to transport energy, so mass flows must be represented. Similarly, cars use energy (from gas) to generate the force for movement, so the energy relationships of the car must also be described.

Generalized Function

Generalized Function explains how the causal laws of the Abstract Function level are achieved. For example, how does energy flow from one piece of equipment to another? Most likely it is carried by something like water or air, or transferred through something like a heat exchanger. When we examine how causal relationships are implemented, we are looking at the processes of the system.

The description of system processes is a more concrete description than that at Abstract Function. For each part of the Abstract Function level, you want to ask "How?" in order to generate the Generalized Function level; for example, "How is a mass source achieved?" Most likely there is a reservoir of water or material. "How does energy flow?" Energy may be transferred through combustion, convection, conduction, or radiation. In fact, if you look at the processes listed in the sidebar, a lot of processes end in the suffix "tion" or "ing." That's just a hint! Most importantly, you want to make sure that this is a different type of description from Abstract Function; ensuring that your language has changed is one way of double-checking.

━━━━━━━━━━━━━ ▼▲▼ ━━━━━━━━━━━━━
KINDS OF PROCESSES

Combustion
Convection
Conduction
Radiation
Evaporation
Condensation
Fission
Distillation
Cracking
Moving
Launching
Digestion
Respiration

━━━━━━━━━━━━━ ▲▼▲ ━━━━━━━━━━━━━

To return to our example of the car, we have decomposed the major processes of the energy flow box. Energy comes in as chemical potential energy through fuel kept in a fuel store. The fuel is oxidized to release heated, high-pressure gas, which performs mechanical work; excess energy is released through exhaust gases and through cooling.

The mass relations could be decomposed similarly. Figure 2.6 shows the Generalized Function level for our car example.

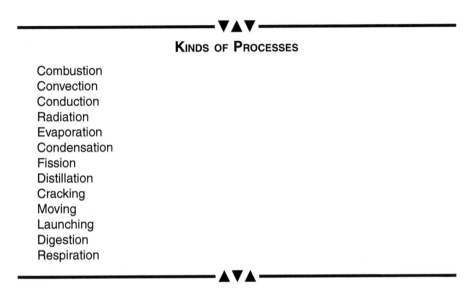

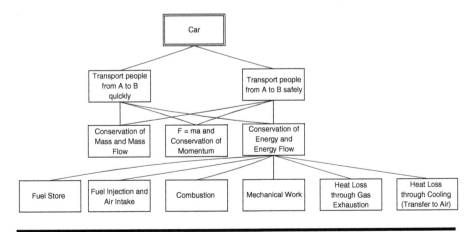

Figure 2.6   Generalized Function level, car example.

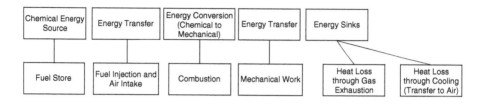

Figure 2.7   Abstract Function and Generalized Function levels, car example, showing more detail of the energy cycle.

From the example you should note certain things:

■ The language is different from the level above. It is more concrete and getting closer to a physical description of how a car works.
■ The entire energy cycle has been mapped out from Fuel Store to Heat Lost to the environment. We could have modeled the Abstract Function in a similar level of detail and identified Energy Stores, Energy Transfers, Energy Conversions, and Energy Sinks. Let's map out the example this way. Our Abstract Function layer (Energy Description) would then look like Figure 2.7.

This is just a more detailed description of the energy flow and conversion from the previous model. The advantage to this description is that the various energy sources and transfers have been described. You will usually want at least this level of detail in your Abstraction Hierarchy models, though as you can see, the model very quickly becomes quite large. It's often useful to create multiple models that show different levels of detail.

Distinguishing Processes from Tasks

Processes and tasks can be quite difficult to distinguish. In both cases, they are sequences of steps required to achieve something. In terms of your work domain model, you want to include processes that the user must monitor. You do not want to include processes that the user must perform with the eventual interface. This dividing line differentiates what needs to be designed at this stage. Examples of tasks that you typically don't want to include in a work domain model are:

■ Monitoring
■ Detecting
■ Managing
■ Reporting

At this point you may be wondering where task analysis fits in with this process. Task analysis can make several important contributions. Task

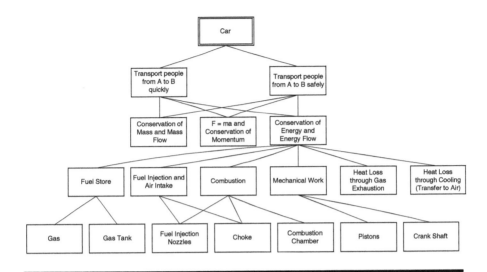

Figure 2.8 Physical Function level, car example.

analysis can help to collect the information that will be represented in the work domain model. Task analysis can be used to improve your EID display further. A good example of combining task analysis and EID is the petrochemical work of Jamieson (2003a, and discussed in Chapter 6). After the EID was built, the interface was improved with task-relevant information, leading to a design that performed better than the EID alone.

Physical Function

When we think of what is in a work domain, we most often think of Physical Function-level information. By Physical Function, we mean the various components of the work domain and their capabilities. If we return to our car example, in the following figure you'll see that this is the level where we describe the gas tank, the engine, and all the common car equipment. You'll notice as well in Figure 2.8 that our example is incomplete, because that's all we had room for.

This level is more than just a listing of components. First you'll notice that each component is used in a process at the Generalized Function level.

PHYSICAL FUNCTION: QUESTIONS TO ASK

What are the components?
What are their capabilities?
How are they involved in various processes?

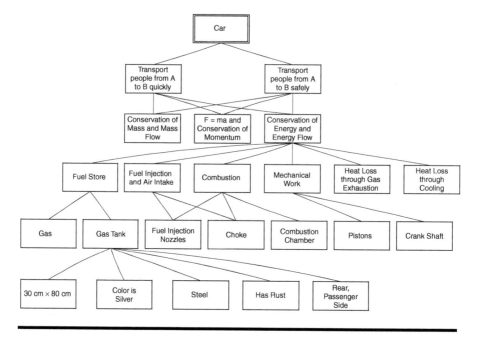

Figure 2.9   Physical Form level, car example.

Second, one of the key areas of interest at this level is capability and, correspondingly, limitations in capability. Therefore, while we list gas tank, we are interested in the capability of the tank to hold gas (its volume), the capability of the engine to generate power, and so on.

Physical Form

Physical Form is a description of the physical appearance of the work domain. This includes the size, shape, color, appearance, location, and condition of components. When working at this level you should think of the information that can be gained from plant drawings, or in-plant video. Figure 2.9 continues our hierarchy by describing Physical Form elements for the gas tank of the car.

Time-Saving Hint: You should develop this level in your model strategically and with a view to how you will use this information. Otherwise, you could find developing this level to be an almost endless task as you determine the color, size, and layout of every component in your system. If you look through the examples in the case studies, you will see that the Physical Form level is rarely done in great detail. That doesn't mean that the analysts don't know what is at that level, but merely that, given the conditions of the project and the eventual design, they have not pursued detail at that level. If needed, the designers can return to the analysis and develop more detail at this level, since they know the information they would need to collect. Physical Form information is useful when you are designing a layout graphic of your domain, creating icons that must look like part of the domain, or making decisions on where to integrate video into the displays. This means learning the layout and connection of equipment, the appearance of major elements, and whether or not in-plant video or drawings will be accessible to the displays.

————————————————— ▼▲▼ —————————————————

ELEMENTS OF PHYSICAL FORM

Size
Shape
Color
Location
Condition
Material

————————————————— ▲▼▲ —————————————————

Functional and Causal Descriptions

The hierarchy built for the car in the previous section demonstrates a functional Abstraction Hierarchy representation. A functional Abstraction Hierarchy representation emphasizes the "means-end" or "how/why" links of the hierarchy. In the five-level tree, only the functional means-end links are shown.

In contrast, a causal Abstraction Hierarchy representation preserves the same structure, but in this case the causal, or "within a level," links are drawn. These links show how the processes and flows are connected to each other within a level. The functional links are removed from the model, although the structure between the levels usually makes the functional links apparent. The two representations are closely related; in most cases, you will want to develop both representations. It is possible to combine them, but separating the functional and causal links often develops clearer models.

Causal links are most useful at the Generalized Function and Abstract Function levels. These levels show flows of materials, processes, mass, or energy, and it is useful to show the sequencing of the flows. To follow the car example we used in the previous section, to develop the causal representation we would remove the functional links and develop the links within each level. Figure 2.10 shows the result of this.

With the causal description, a different set of relationships can be seen. For example, the causal description clearly indicates that the combustion

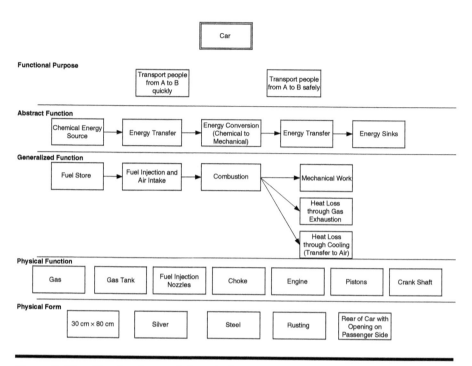

Figure 2.10   Five-level causal description, car example.

process results in mechanical work and heat losses simultaneously. The causal description also shows the flow between the Abstract Function and Generalized Function elements.

──────────────── ▼▲▼ ────────────────

FOCUS ON TERMS

Functional Abstraction Hierarchy — An Abstraction Hierarchy that explicitly shows means-end links
Causal Abstraction Hierarchy — An Abstraction Hierarchy that shows flows within each level

──────────────── ▲▼▲ ────────────────

PART-WHOLE HIERARCHIES

You can see from the car example that Abstraction Hierarchies quickly become fairly detailed. In particular, when we develop hierarchies for large systems, we develop them in various levels of detail in what are called Part-Whole Hierarchies. A Part-Whole Hierarchy is actually a second kind of hierarchy, one built on aggregating components.

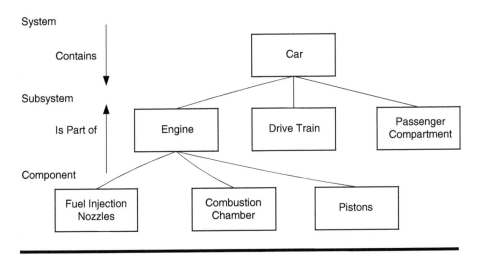

Figure 2.11   Part-Whole Hierarchy, car example.

Figure 2.11 shows a Part-Whole Hierarchy for the car example that we have been using in this chapter. At the top of the hierarchy the car is described as a whole, or a "black box." It is just listed and not detailed at all, and all the components are aggregated in the single description of "car." At the next level, major subsystems within the car are broken down; in this case three subsystems are shown: engine, drive train, passenger compartment. At the final level, one of the subsystems, in this case the engine, has been broken down into its components: fuel injection nozzles, pistons, and so on. You will notice that the guidelines on building a hierarchy have been maintained. The structure is tree-like, there are different levels, and the levels are ordered one above the other using a consistent relationship. In this case, language used for the relationship is "contains" going down and "is part of" going up.

Unlike the Abstraction Hierarchy, Part-Whole Hierarchies do not have a set number of levels. You can adjust the number of levels in the hierarchy depending on the complexity of the work domain that you are analyzing.

When we do a WDA, we use both Abstraction and Part-Whole Hierarchies together in the analysis. We generally put the Abstraction Hierarchy on the vertical axis and the Part-Whole Hierarchy on the horizontal axis to form a two-dimensional space, as shown in Figure 2.12.

If you compared our Part-Whole Hierarchy of the car with the Abstraction Hierarchy, you may have noticed a major similarity with the analyses. If we look at the Part-Whole Hierarchy, it starts at the level of car and then moves down to components. Similarly if you look at the Abstraction Hierarchy in Figure 2.10, you will see the hierarchy discusses the car at the Functional Purpose level, and components at Physical Function. The

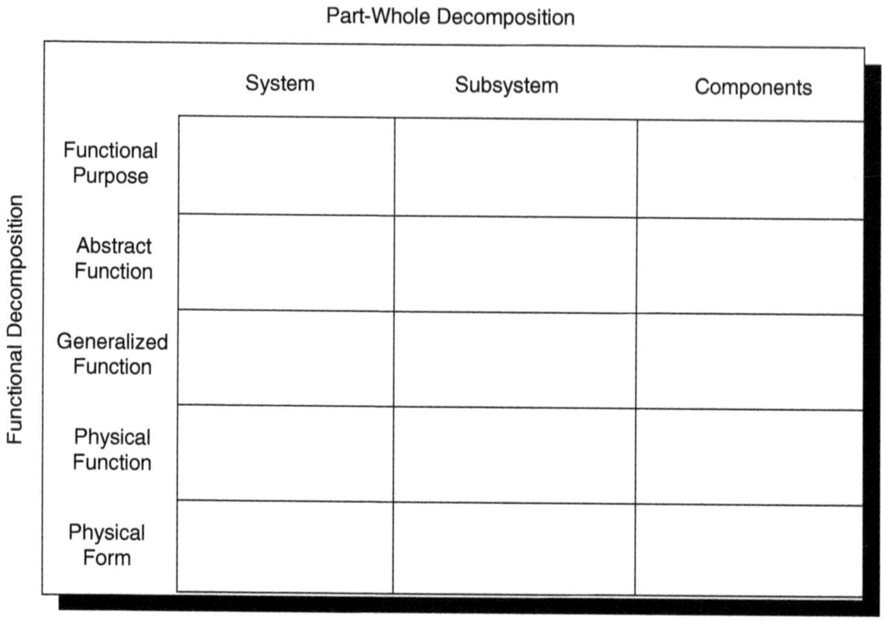

Part-Whole Decomposition

	System	Subsystem	Components
Functional Purpose			
Abstract Function			
Generalized Function			
Physical Function			
Physical Form			

Functional Decomposition (vertical axis label)

Figure 2.12   Combining Abstraction Hierarchy and Part-Whole Hierarchy to make a work domain model.

Abstraction Hierarchy itself has some degree of Part-Whole Decomposition in it, though certainly not as explicitly as in the Part-Whole Hierarchy. This is because Functional Purposes tend to relate to the work domain as a whole, and work domain capabilities (at Physical Function) are best described at levels of components. Although we have drawn the two dimensions as orthogonal in Figures 2.12 and 2.13, in reality they are not completely orthogonal. Some cells of the matrix generate a more useful description than other cells. For example, while it makes sense to discuss the purpose of the car as a whole, describing the purpose of individual components is not very useful.

Figure 2.13 gives you an idea of what this description may look like. It was developed using both Part-Whole and Abstraction Hierarchies. Some elements, such as the mass flows, have been left out to keep the figure simpler. Notice that down the first column we look at the car as a whole, in the middle column we look at just the powertrain subsystem, and in the final column we look at just the engine. To complete this model, you would need to develop the component level for all components in the powertrain. Also, all the subsystems and their components would need to be developed.

Part-Whole Decomposition

	System	Powertrain	Engine
Functional Purpose	Transport people from A to B quickly and safely		
Abstract Function		Energy Source Energy Conversion Energy Sinks	
Generalized Function		Fuel Source Combustion Heat Exhaust	Fuel Injection Combustion Heat Exhaust
Physical Function		Fuel, Gas Tank, Fuel Injectors, Engine, Exhaust Components	Engine Components
Physical Form		Location, Size and Appearance of Powertrain	Location, Size and Appearance of Engine

Functional Decomposition (vertical axis label)

Figure 2.13   Work domain model, car example.

One of the reasons we make a Part-Whole Hierarchy is that it gives us insight into how to group displays. The system or subsystem levels indicate information that should be on an overview or status display. The component level indicates the information needed for a more detailed display or a display that operators would control from.

MODELS WITH SOCIAL CONSTRAINTS

There are domains that are constrained by more than just physical constraints. In some cases, social rules, values, and priorities present legitimate constraints on action. In these domains, it is important to include these aspects in your work domain model. An example of a domain with a mixture of physical and social constraints is the military domain. Although military equipment has certain capabilities, the application of this equipment is heavily constrained by the values, priorities, and rules of engagement that cascade down from higher command levels to the individual user. In this domain, working outside these constraints is almost unheard of, and interviews with users will quickly reveal how important these constraints are.

Social constraints can be modeled with WDA in roughly the same style as physical constraints. By this we mean that social constraints have a purpose, principles, processes, and often some kind of physical instantiation. Figure 2.14 shows a hierarchy with social constraints.

You will note that the typical labels for the analytical levels are a clumsy fit to the social constraints. There may be more relevant labels. For a case that includes social constraints in the analysis, see the example of the Halifax Class frigate in Chapter 5.

TECHNIQUES FOR MANAGING LARGE MODELS

In modeling a large system, work domain models quickly become unwieldy. We have used two techniques to help organize the models: systematic decomposition and link tables.

Systematic Decomposition

Systematic decomposition is nothing more than an organized way of approaching a large work domain model. We used this approach on a project with Syncrude Canada, where the plant had dozens of subplants, all with hundreds if not thousands of components. We also had seven analysts working on the models, so maintaining consistency across the work was very important.

In this project we began with the Part-Whole Hierarchy of the plant. We had models from the overall black-box model of the plant itself to major subsystems. Figure 2.15 shows the structure of the Part-Whole Hierarchy.

Expanding consistency across all plants, each Part-Whole Decomposition was done on a separate page. This decomposition structure created the foundation for the Abstraction Hierarchy models. We followed the same approach, with a work domain model for each page, for example, Syncrude (overall), Syncrude plants, Plant y, Plant y subsystem 1. Doing the Part-Whole Hierarchy first was instrumental in helping us to structure the overall work domain model. See Chapter 6 for more detail on this project.

Link Tables

Another problem with large projects is that it is easy to forget the rationale behind the various links in the work domain models. Inevitably, a client will identify a particular link and ask you why it is there. If it has been a long time since you created the model, it is easy to forget the reason.

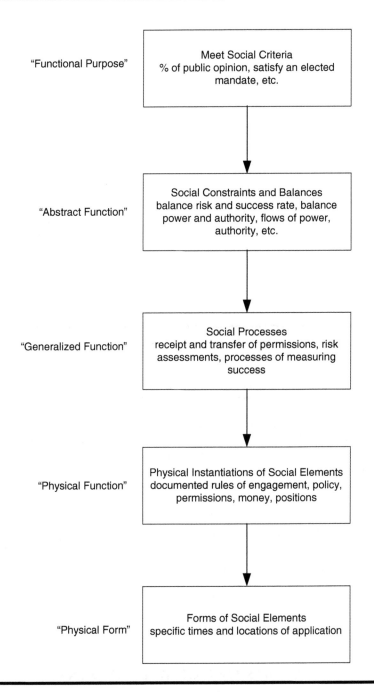

Figure 2.14   Work domain model with social constraints.

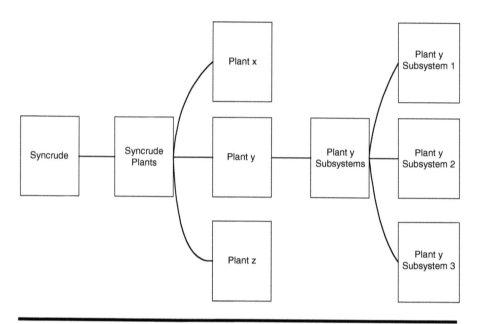

Figure 2.15   Structure for a large modeling exercise.

Table 2.2   Recording Links between Levels

GF	AF	Rationale
Water source	Mass source	The water tank is the primary mass source
Emergency water source	Mass source	They can switch to this tank if needed

You may develop your work domain model with a tool that allows you to label the links; this is a good idea. Another technique we have used in the past is to record the link rationales as you develop them in a table. You'll need four tables, one for each of the link regions in the model (AF to FP, GF to AF, PFn to GF, and Pfo to PFn). Table 2.2 is easily constructed by listing various elements.

The key to managing large models is being systematic and organized, and documenting your model as you do it.

MODELS OF MULTIPLE DOMAINS

While a Part-Whole Decomposition is one way of parsing a large domain, there are cases where a domain should be parsed for another reason. Some domains have distinct regions of control that should be differentiated.

Table 2.3   Multiple Domain Model for a Business Domain

Level	Business	Supplier
FP	Maximize profit Minimize supply costs	Maximize supplier profit
AF	Resources in, product out, $ flow	Supplies out, $ in
GF	Manufacturing processes, ordering	Supply/delivery processes
PFn	Supply, product, staff, etc.	Supply
PFo	Location	Location

Sometimes the users do not have full control over all regions of their work domain. One example of this situation is ambulance dispatching. While the dispatcher has control over the various ambulances and personnel, the dispatcher has limited to no control over the injured party. The injured party is still part of the dispatcher's work domain, since they must match ambulances, personnel, and equipment to the various needs of the injured people. Another situation with two control regions is aircraft navigation. While the pilots must operate the airplane, the airplane is exposed to environmental conditions and landing and airport conditions, which are largely out of their control. In this case the environment is still part of the work domain, although it is a distinctly different region of control than the plane itself.

In these cases with different regions of control, it is important to identify this in the work domain model. The easiest way to do this is to make a separate work domain model for each region. You'll want to keep track of various interconnections between the two models, though, since these interactions are often the richest sources of work domain information.

The simplest way to do this is to divide the work domain model but still allow crossover between the two domains, as shown in Table 2.3.

As you can see in this model, supply is important in both work domains, while only the supplier has control over it. The business, however, communicates with the supplier through its ordering processes, and the two domains come together at the lower two levels, where the quality, type, and location of the supply intersect. Another important interaction, though in this case a conflict, occurs at the highest level. The business wants to obtain supplies at the lowest possible cost, whereas the supplier wants to maximize its return on the supplies. The two-level model has provided added richness over a business-only model. If we include all of these elements within the same model, we leave something very important out. We have not clearly differentiated that some elements of the supply are out of the control of the business. Showing your user what can and what cannot be controlled is always an important distinction.

There are other variations of split models that demonstrate the flexibility and advantages of using these kinds of models. Unlike the earlier business example, sometimes we have parties with similar goals but different regions of control. The ambulance dispatching domain is an example of this. In this case there may be little similarity at the lower levels, and higher congruency at the top levels.

TESTING YOUR MODEL FOR COMPLETENESS

Testing a work domain model in a formal sense is a difficult endeavor. A work domain model is a representation of the world; while it needs to be accurate, how detailed the model is and the form of the representation can vary widely. Two models of similar domains can have differences in both representation and content. For an example of two work domain models of similar domains, see Bisantz et al. (2002). Like any other model, the value of a work domain model is for the most part justified by how useful it is in later activities. A good model will generate new kinds of information and relationships for display design. A poor model will not. Usefulness is the ultimate test of a work domain model.

There are situations, though, where some form of model validation can be a useful exercise. When analysts are modeling an unfamiliar domain, a validation exercise can confirm existing information and sometimes lead to new insights. Some clients may also be more comfortable knowing that the work domain model has been taken back to domain experts for confirmation. If the validation is performed properly and interpreted wisely, then it can be a very positive stage in the design process.

The objectives of a validation of a work domain should be as follows:

1. Confirm that the relationships in the model exist and connect in the manner shown.
2. Confirm that, within the scope chosen, the model is not missing information or relationships.

In most cases, the validation exercise should not:

1. Become an argument over representation, and what element is at what level. The representation should remain the decision of the modeler. Small differences in representation will most likely not affect the usefulness of the model, providing that the model is accurate and complete.
2. Become an argument over the scope of the problem space. The scope of the problem space should have been decided before the modeling began.

3. Become an exercise in prioritizing or weighting the elements of the model. Work domain modeling is a broad modeling exercise, and, at this stage, there is little to be gained by prioritizing the different elements of the model (e.g., ranking by frequency used or critical-ity). A particular physical constraint or little-used piece of equipment can have a dramatic role in an unexpected accident situation. A frequency-of-use prioritization could lead to a limiting design. A more effective approach would be to design your EID display and apply task analysis to optimize certain task performance.

These are, in our opinion, the three major traps of performing a validation exercise on your work domain model. However, there are several useful techniques you can use to gain more confidence in your model. Two techniques that we will discuss are scenario mapping, and questionnaires or surveys developed from the work domain model.

Scenario Mapping

Scenario mapping is a useful technique that can confirm work domain models, obtain new information, and provide a working demonstration of the models. Scenario mapping can also take advantage of pre-existing materials, such as training scenarios, design basis scenarios, and task analysis information. Scenario mapping first appeared in the literature in 2001 (Burns et al. 2001). Chapter 5 gives this particular example in detail. In scenario mapping, a scenario is "walked through" while checking that the work domain model contains the requirements to complete the scenario.

The key considerations for conducting this kind of exercise are as follows:

1. The work domain models must be created before the exercise. It is important that the work domain models maintain a breadth of perspective beyond the scenario set.
2. You need one or more scenarios that provide good coverage of the work domain. You should cover the work domain in physical space (i.e., make use of different components) and in complexity and interactions.
3. You need domain experts with operational experience. Like most walkthrough evaluations, a large number of domain experts are not required; three to five experts will identify most problems.
4. You should allow the domain experts to work in their own language and world, rather than forcing your domain experts to learn WDA. This means the analyst must prompt and probe for information,

and provide the translation from operational concerns to the work domain model.

5. The test can generate missing information, but by no means can be considered a complete or formal validation. Only the information that is called for in the scenario set is being tested.

As a general procedure, we recommend the following steps:

1. Make multiple copies of your work domain model for your own note-taking.
2. Have domain experts step through the scenario. At each stage in the scenario, probe for work domain-related information. Examples of the kinds of questions you might ask are:
   ■ What equipment are you using at this step?
   ■ What are your considerations in using this equipment? Is anything limiting its use or helping it to work better?
   ■ If that piece of equipment was unavailable, what else could you use?
   ■ Why are you using that equipment? What are you trying to achieve?
   ■ How can you tell if it is working correctly?
   Note that none of these questions refers to the work domain levels. The emphasis is on working with the domain experts in their domain, the operational domain, and translating the results later.
3. Translate your notes onto the work domain model. Start by indicating the key pieces of equipment at that stage and then indicate the higher level functions. Keep in mind that whenever a piece of equipment is being used, its higher level functions come into play. At each stage, therefore, you should have a map that covers all levels of the work domain model.
4. Develop a series of work domain pictures that show how the work domain is used across the scenario. Figures 2.16 and 2.17 show examples of these work domain pictures.
5. If new information arises from the exercise, revise your work domain model accordingly.

In Figure 2.16 and 2.17, different regions of the work domain model are highlighted. In each case, the highlighted regions are completely connected; that is, the higher level functions are indicated as well. Figure 2.17 gives an example of finding a missing link through the exercise. The model should therefore be revised to reflect this correction.

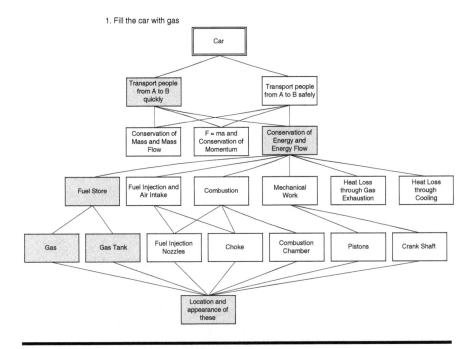

Figure 2.16   One snapshot from a scenario mapping exercise.

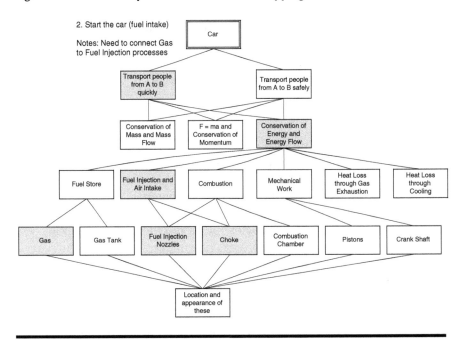

Figure 2.17   Next stage in the exercise.

Testing Your Model through Questionnaires

Testing the model through questionnaires is more difficult than scenario mapping, because questionnaires are rarely as rich in contextual information as scenarios. It is possible, however, given the right set of conditions and interpretations.

The general approach in testing through this method is to translate your work domain model into a set of questions that can be evaluated. In the past we have used a straightforward translation, where each model element is listed and participants in the evaluation are asked to rate whether the element comes into play "Never," "Sometimes," or "Often." Participants can also be asked if there are elements that they consider to be missing and invited to list those elements. An example of these basic model testing questions are:

1. In the car, the choke is used   Never   Sometimes   Often
2. Gas is used   Never   Sometimes   Often

As you will see from the example, a very thorough approach that works through each model element can result in dull or even embarrassingly simple questions (Question 2, "Gas is used"). You may find it more useful to limit this exercise to some key areas of your model.

You can also probe for connections in your model, and this can make for more interesting questions. In general, you want to set questions that probe down a level (Question 3) and up a level (Question 4). For example:

3. In balancing the fuel:air ratio, the car uses:
   x Gas
   x Air intake valves
   x Choke
   x Fuel injection nozzles
   x Other equipment not mentioned here _____ air intake fan _____
4. Fuel injection is used to:
   a. Transport gas into the combustion chamber
   b. Establish adequate combustion chamber pressure
   c. Aerosolize the gas particles for more efficient combustion
   d. Any other purposes?_____

Again, just like in the scenario mapping exercise, the emphasis should be on looking for new information and missing relationships. The questions should not be used to prioritize elements. In order to get reasonable results, you need to word the questions operationally and then translate them to your model later. Do not ask representational questions (e.g., "Is fuel injection a process or an Abstract Function?"). It is your job as the analyst to understand how to represent the work domain. For an example of testing a work domain model, more detail is provided in the frigate model in Chapter 5.

SUMMARY: DOUBLE-CHECKING YOUR WORK DOMAIN MODEL — RULES TO USE AS YOU ACTUALLY BUILD IT

Now that you understand the levels of the Abstraction Hierarchy and different ways of using it, there are some simple rules that will make it easier to actually develop a work domain model. We will go through these step-by-step and summarize the rules at the end of the chapter.

Define the System of Interest

Before you start, you should make sure you understand exactly what system you are analyzing. Where does the system start? Where does it end? What do you have to include? This is analogous to most other engineering analyses, where you begin solving the problem by defining the system boundary. With a system boundary, your analysis is scoped and clear. Without it, you may not know when to end your analysis, or you may not be sure that you have covered everything. Defining the boundary helps to scale the analysis and also to identify what is inside or outside of the work domain. Depending on the purpose of your analysis, you may have different system boundaries. For example, if you were analyzing the work domain of nuclear power generation, there are several potential boundaries. You could look at power generation across a country and how various plants contribute to a nation's energy resources. You could look at generation in a single plant, drawing the boundary at where the plant connects to the external electrical grid. Alternatively, within the plant you could analyze a single unit and look at how that unit performs. Where you draw the boundary can significantly change the nature of the problem. Consider the differences at the levels of Functional Purpose and Physical Function for each of the three different problem spaces shown in Table 2.4.

The purposes of the work domains are different, as are the basic components of the domains. Predictably, operation at each of these levels would have different challenges. Whereas unit operators may be working towards maximizing their individual unit's output, the plant operators may be trying to manage the output of various units to meet a predetermined energy requirement for the electrical grid. A countrywide power management operator would be concerned with grid performance across the country and maintaining energy distribution levels to various areas. That's

Table 2.4 Model Differences with Different System Boundaries

Abstraction Level	Countrywide	Plant	Unit
Functional purpose	Meet national energy requirement	Produce energy (utility set level)	Meet designed energy production for unit
Physical function	Plants, electrical grid, energy users	Units, storage, internal grid	Generator, turbines, etc.

why your first task, even before you begin building your Abstraction Hierarchy, is to define your system boundary.

If you are unclear about your system boundary, we suggest that you err on the side of breadth. You can always define the system broadly and then work out the details as you work through your Part-Whole Decomposition.

Start Building Your Abstraction Hierarchy from the Top

The first challenge in understanding your work domain is to determine what it was designed to do. Determining this kind of purpose actually constrains the rest of your analysis, so it makes sense to start your analysis here. Functional Purpose can often be determined by taking a bird's eye view of the system and doesn't require detailed knowledge of the system to complete. Just as a reminder, when defining Functional Purpose you want to look for at least two purposes. Likely one purpose will be performance related. The other purpose may be something like safety or efficiency. You want to describe your purpose with adverbs or a quantitative description of what would constitute good performance. Consequently, you should be able to define poor performance by looking at your Functional Purpose level.

Next Work from the Bottom

After defining your Functional Purpose, we suggest that you begin working on the Physical Function and Physical Form levels. These levels are the most concrete levels and, therefore, are easy to understand. Determining your components can also help you decide whether or not you need to do a complete work domain model and work your Abstraction Hierarchy along a dimension of Part-Whole Decomposition.

The easiest way to complete these levels is to begin by listing all the components of the system and then, in Physical Form, defining the location and appearance of the components. When you have made your list, check it for two things: Make sure the current interface (keyboard, screen, icons, mouse, displays) isn't included in the system. Because you are doing this analysis to build a new interface, you don't want to include the elements of the old interface.

The other kind of components you usually don't want to include are controllers and control systems. For the same reasons that we leave out the interface, if we leave out the controllers, we have a chance to redesign them or to change the allocation between the human operators and the controllers. If, however, the control systems cannot be changed and the operators must monitor them, then it is best to include the control systems in your analysis.

Complete the Middle

By this stage you should have your Functional Purpose, Physical Function, and Physical Form levels completed, or almost 60% of your analysis. There are only the two middle levels to do. Keep in mind that Abstract Function describes causal laws and conservation principles. Generalized Function, however, is more concrete and describes the processes that achieve those causal relationships.

Make Sure Every Box Is Connected Up and Down

Once you have filled in all the levels, review your hierarchy. Every box should have at least one connection to something directly above it and to something directly below it, aside from the anchor levels. If a box is unconnected, something is wrong with your analysis. Ask the following questions:

1. Do I need to add connections to the box?
2. Is the box at the right level?
3. Could something be missing from either the level above or the level below?
4. Has the box already been described elsewhere in the model?

Check Your Language

Keep in mind that one of the keys to building a hierarchy is that every level is different. You should never have identical or near-identical elements at different levels. Similarly, at each level there should be consistency in the language of description. If there isn't, chances are some of your elements are at the wrong level. The language will give you a clue about how to correct it.

Develop Detail as Time Allows

A WDA can be a lengthy analysis to complete. If you are working under time constraints, it will be more worthwhile to complete a broader analysis, with less detail, than a detailed analysis of a subsystem. The reason for this is that the broader analysis shows the interconnections between the different subsystems. It also gives you a framework to expand your analysis if more time becomes available in the future. Even at a high level of detail, a work domain model can be a very useful design tool.

Translate Your Hierarchy in Variables, Constraints, and Relationships

This is the topic of Chapter 4.

Show the Value of Your Analysis

You have just done a lengthy analysis using a relatively new analytical approach. Your final step, therefore, should be to determine what you have gained from the analysis. From experience, in most cases you gain variables at the Abstract and Generalized Function levels, as well as a keener sense of the key relationships. So, at this point you need to compare your list of variables with the variables on the current display. Are there any new ones? Have you learned any new relationships? If the answer is yes, then you are ready to begin designing an EID.

▼▲▼

9 HINTS FOR A SUCCESSFUL WORK DOMAIN ANALYSIS

1. Define the system of interest.
   Err on breadth.
   Draw a system boundary.
2. Start at the top. Determine purposes.
   What was the system designed to do?
   How could you define "good" performance?
   What else is important besides performance? Safety, efficiency, profit, environment?
3. Next work from the bottom.
   List all available resources and equipment.
   Location, appearance, etc.
   Make sure the "interface" is not included here.
   Make sure controllers are not listed.
4. Complete the middle.
5. Make sure that every box is connected up and down. Resolve any unconnected boxes. Unconnected = something missing, something extra, or something in the wrong spot.

6. Check language. Each level is a different kind of description of the system and uses its own language.
7. Develop detail as time for analysis allows.
8. Translate into variables, constraints, and trade-off relationships.
9. Show value. What have you learned that isn't in the current interface?

3

THE LANGUAGE OF INTERFACE DESIGN

Before designing interfaces, we need to establish a common language for discussing interfaces. In this way, we can explore, critique, and describe interfaces more consistently.

Although finished interfaces are perceived as integral objects, interfaces are created from many levels of nested designed objects. At the lowest level we have the design of individual pixels, and at the highest level we have large interface structures of multiple displays. The amount of freedom that is possible in working on a virtual display surface magnifies the number of design decisions that need to be made and the number of errors that can occur in designing interfaces.

INTERFACE DESCRIPTION

Woods (1997) has proposed a useful hierarchy of interface objects that we will adapt for this discussion. This hierarchy is as follows:

- Workspace
- Views
- Graphic forms
- Graphic fragments
- Graphic atoms
- Pixels

Each of these levels builds on the level below it and to a certain extent is constrained and dependent on lower level design decisions. The relationships are nested relationships, with workspaces being composed of

Table 3.1   Design Decisions at Different Levels of Visual Form

Level	Example	Description	Decisions
Pixels	.	The smallest graphical unit, constrained by the limits of the screen	"Color" or light emission
Graphic atoms	A, 3 __	Composed of pixels; a letter, digit, line, or color block	Color, size, shape, thickness, angle, forms of reference
Graphic fragments	Word, 2002, scale	Composed of graphic fragments; a word, number, scale; not a complete graphic form	Position, content, organization of fragments, forms of reference, proportion and salience
Graphic form	A graph or indicator	Composed of graphic fragments, this level conveys meaning	Analog and digital forms, display of context, salience across graphic fragments
Views	A window or single cohesive display screen	Composed of graphic forms, this level brings related graphic forms together to describe a process or show sequence	Relations across graphic forms, salience between forms, organization of forms
Workspace	The entire display application	Composed of views, the workspace defines the virtual action space of the operator	Relations across views, navigation, overview, and workspace status

views that are composed of graphic forms and so on. Table 3.1 describes each level and the design decisions that are typically made at that level.

There are, of course, workspaces that are comprised of only a single view, and views that are only a single graphic form. However, for the most part, this is a useful structure to begin a discussion of computer interfaces.

You should note that the design decisions at each level move from elemental decisions to relational decisions as we move from low-level graphic elements to higher level workspace design. While at the lowest levels (graphic atoms and graphic fragments), there are many design decisions related to form, color, and positioning, at the higher levels, the issues are relational issues such as salience and organization. This brings out an important aspect of interface design. Interface design develops relationships between designed objects. Adding a new object to an interface immediately changes how all the other objects will be perceived. Many times this requires the first set of objects to be redesigned or modified. The design process is almost inevitably cyclic.

In the next sections, we will discuss elements of these design decisions, focusing on the choice of forms of reference, analog and digital forms, context, and salience.

FORMS OF REFERENCE

One of the earliest decisions, arising even at the level of graphic atoms, is the decision on how to refer to a piece of data. There are three different alternatives: You can refer to it propositionally, iconically, or analogically. These categories are derived from semiotics, or the study of communication. Woods (1997) applied these categories to interface design, and we'll continue to use them in our discussion. Choosing between these alternatives establishes a particular type of communication between your user and your data. One form of communication is not better than another form; the best interfaces use all forms effectively.

Propositional Forms

Propositional forms tell your user about the data. We are very familiar with propositional forms: Written and spoken language is the most common example of a propositional form. Propositional forms are built upon a commonly understood set of symbols or representations. These representations do not need to look like the data, but the conventions governing these representations need to be understood. As an example, written language conveys a lot of information based on a symbolic set of 26 marks (or letters) organized by convention into words and a grammar or language. But this information is only accessible to people who understand these rules. Written language, therefore, is a poor choice in situations where people may not know how to read, or people may read a different language.

Propositional forms, like languages, are to a certain extent arbitrary. They are learned through experience and adopted by convention. For the user who is experienced in the culture, the convention becomes natural and automatic. However, for novice users this convention relies on their memory, and they may have to memorize certain conventions when learning your system. Propositional forms, therefore, may not be a good form to use in situations where memory may be overloaded or you must design for new users.

Iconic Forms

Iconic forms differ from propositional forms because they look like the data or the object that they represent. We see iconic forms frequently in computer software and on signage. They are used on signage particularly because they do not depend on language and therefore are a good choice in an environment where people may speak different languages. Examples of this kind of icon are the airplane arrival and departure icons used at airports. Airports require signage that doesn't rely on language in order to help a diverse population of travelers. For example, Transport Canada uses the pictographs in Figure 3.1 for arrivals and departures.

For the most part, these pictographs portray airplanes either taking off or landing. An astute observer, though, will note that planes take off and land mostly horizontally; this dimension has been exaggerated to develop the icon. Note also how the arrivals icon even shows the landing gear of the plane.

Iconic forms often rely on users having experience with the domain, otherwise they may not recognize the object. Iconic forms also rely on the skill of the designer in developing a visual form that truly looks like the object. Otherwise the visual form may not be as iconic as intended; it may really be propositional. An example of a propositional icon is the emergency icon on cars, shown in Figure 3.2. The icon uses a symbol that

Figure 3.1   Icons used by Transport Canada.

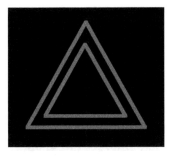

Figure 3.2   Emergency icon used in cars.

users must become familiar with before they recognize it. This reference is propositional and, though graphical, is relying on learned information and stored rules. Whereas propositional forms rely on recollection of a set of stored rules, iconic forms when well executed evoke recognition.

ARE ALL "ICONS" ICONIC?

Just because we call it an icon does not mean that the visual form is truly iconic. The word "icon" has been adopted in computer interface design and used to describe any small graphic symbol that represents something. Many "icons" are actually propositional forms, meaning that they are symbols for something and don't actually look like the action or information. Can you think of some examples of this?

Analogical Forms

Analogical forms capture some sort of constraint in the environment and map that constraint onto the visual form. A good example of an analogical form is a map, which takes the spatial relationships and distances of the real world and transfers them using a scaling relationship into similar relations on a piece of paper. Some analogical forms may look like the information that they represent and therefore capture iconic properties as well, but this is not necessarily true.

Conveying Information with Different Forms

Conveying information can be easier or harder depending on which form you choose. It is often difficult to convey context and changes in data with propositional forms. Consider the following small design example:

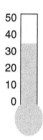

50
40
30
20
10
0

Figure 3.3   Basic analog display of a thermometer.

> The temperature today is 5 degrees Celsius, dropping to 0 by the end of the day.

The description of the example itself is a propositional reference about a state and prediction of weather in the environment.

If we wanted to design an iconic form for this information we would immediately realize that "5 degrees" has no real visual representation. However, a thermometer, Figure 3.3, is a common measure of temperature and might be used as an icon for this situation. We would still have difficulty conveying the future drop in temperature on the thermometer.

A thermometer is also an analog representation, where the change in energy level of the environment has been mapped to vertical movements on a scale. The original constraint was created physically by the expansion of mercury or alcohol constrained in a thin tube placed against a scale. This is a very good example of transferring constraints from the environment into a visual form. In order to show the change in temperature, however, there may be other more effective analog forms. For example, a trend chart showing temperature over time, as in Figure 3.4, would show the expected decrease easily.

Analog forms are often most effective when we need to convey changes in data or many relationships in a concise format. The temperature example is fairly simple, but if we also wanted to convey normal temperatures for that day, such as expected highs and lows, you can quickly see that using a propositional form would require a short paragraph, such as a weather report, to convey all the information. Similarly, the iconic display would begin to require a collection of icons. In contrast, this richer information is easily shown on the trend chart in Figure 3.5.

Analog and Digital Forms

Many of the displays that we see in modern industrial systems are either analog or digital graphical forms. These are implementations of the forms

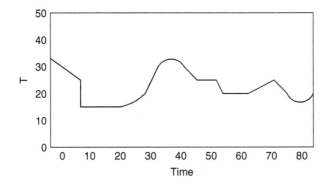

Figure 3.4   Basic trend display of temperature vs. time.

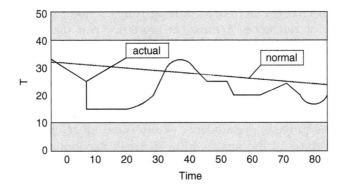

Figure 3.5   Trend chart of temperature vs. time with normal, high, and low information.

of reference discussed in the previous section, with digital forms being primarily propositional forms of reference and analog forms using analogical forms of reference. It is important to understand how to use these forms effectively.

Digital Forms

Digital forms are alphanumeric representations of data. They are effective where precision is required; they tell the user the exact measured value at the point in time. In fact, in modern computer systems, determining the level of precision that is needed, or that is even reasonable given normal amounts of sensor and calculation error, is a design decision. While digital forms are very precise, it can be difficult to display information over time or context with digital forms. If these are not shown, the user must remember these values.

Table 3.2   Properties of Analog and Digital Graphic Forms

Characteristic	Analog	Digital
Precise reading or recording	Poor	Excellent
Showing context	Excellent	Challenging
Showing history	Excellent	Challenging
Saving space	Poor	Excellent
Form of reference	Analogical	Propositional

Analog Forms

Analog forms are graphical representations of data. While not as effective at supporting precise reading, analog forms can be effective for showing context and changes over time. Analog forms require the establishment of a frame of reference. A frame of reference is a background for the data. This frame of reference determines how easily changes in the data can be observed and sets the context for the data. Establishing the right frame of reference is the best way to make data informative.

Analog forms can have one- or two-dimensional frames of reference. A trend chart is an example of an analog form with a 2D frame of reference. Selecting the range and granularity of this frame of reference can impact the effectiveness of the analog form. If small variations in the data are critical, you need to use a narrow range on your scale. If larger changes in the data are important, a wider range should be used on the scale.

Analog and digital forms can often be combined in order to take advantage of their different benefits (Table 3.2). An example would be pairing a bar graph with a digital readout.

Context

When we are providing context in a display, we are providing background information for data. This background information allows us to determine if the data is:

- Average, high, low, or off-scale
- Normal or abnormal
- Increasing or decreasing
- Approaching alarm limits or setpoints
- Approaching a point where action must be taken

Context is something that we always search for when we receive any kind of information, computer display or not. If you get an exam returned with a mark of 27, you will want to know:

- What were the total possible marks?
- What was the class average, or standard deviation?
- What was the range, or highest and lowest marks?
- What was the exam worth as a percentage of the course?

These are all elements of putting the data into context, or rather, turning it into information. Consider the different information value of receiving a grade of 27 out of 30 versus 27 out of 100. Context is what makes it meaningful.

THE CONTEXT CHECKLIST

How does your design show
High, low, mid-range, off-scale
Normal, abnormal
Maximum, minimum values
Increasing, decreasing
Average, standard deviation, or expected variability
Alarm limits, setpoints, action points

Salience

When we design ecological displays, we often provide a lot of information to our users. Controlling what they pay attention to is a key way to manage this large amount of information. Designing salience is about designing what "sticks out" in an interface. When we design salience we make decisions about what information our users need to see first, second, third, and so forth. The careful design of these relationships allows us to put more information on a display, with less cost to the user in terms of mental effort.

Salience is affected by several design choices. Objects are more salient if they are:

- Large
- Moving
- Flashing
- High contrast with the background

Note that we have said high contrast with the background rather than identifying specific colors. While it is true that yellow and red are often used as high-salience colors, salience is a truly relative quality. Consider the following small example in Figure 3.6. In the left drawing, a large

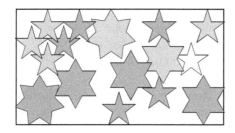

Figure 3.6   Salience is relative.

gray star is a salient object. The salience relationship is reversed in the right drawing, with the only change being the change in the background. Now a small white star is salient. Because salience levels are relative, salience changes with each new addition to the display.

A VISUAL THESAURUS FOR DATA RELATIONSHIPS

Ultimately, when we are designing an ecological display, we want our users to be able to perceive relationships. We use the WDA to start identifying those relationships, but the final step is to communicate those relations in a visual form. In this section we discuss certain common relationships that develop from a WDA and a generally effective means of displaying those relationships. The relationships can be categorized as single variable relationships, multivariate relationships, and structural relationships. Within these categories we discuss:

- Single variable display options
  - A variable within limits
  - A variable with a constraint
  - A variable with a normal value
  - A variable that changes with time, or where the rate of change of the variable is of interest
- Multivariate display options
  - Variable balance, or variable = variable
  - Variables are additive
  - Variables are multiplicative
  - Multiple variables determine system state
  - Multiple variable balance
  - Multiple variables with interacting constraints
- Structural display options
  - Linear structures
  - Spatial structures
  - Symbolic structures

These display concepts are not the only alternatives, but rather a guide to some fundamental elements. These concepts reappear in the case studies later in the book.

Single Variable Display Options

Variable within Limits

In almost all displays, many variables will have to be shown. The variable within limits, therefore, is the most basic type of visual display. All the other display concepts discussed in this section are enhancements of this basic idea.

Whenever we identify a variable or a value that needs to be shown, we need to determine what is important about that variable. What is a high level? What is a low reading? What are normal fluctuations? If this sounds familiar, it is — these are the basic elements of establishing context by determining a frame of reference.

In this basic form of visual display, the frame of reference is usually the expected range of the variable itself. This range may be determined by the range of the sensor, or, if the sensor range is very large, you may need to reduce the scale you are using. The most critical parts of your scale, or your frame of reference, are showing normal, too high, and too low.

In most cases, an analog form is appropriate for this kind of information. If your users are interested in knowing if the variable is on target or ranging too high, the analog form should be sufficient. The analog form is useful because it shows the user how close the value is getting to its limit. A value that is far away from a "too high" limit has a totally different meaning from a value that is just under the limit.

If your users need to monitor for a specific value, or report or record values for other users, you should add a digital display to your graphic. It is also possible to show high and low values with a digital display. Sometimes this is a good choice in tightly constrained display areas.

Here are some possible designs for showing a variable ranging between high and low limits.

Bar Graph

Characteristics: The current value is shown against a frame of reference of its range (Figure 3.7). High and low limits are easily shown. The distance to a limit is visible, so this is useful if proximity to a limit is something users want to know. The graphic can take up a lot of space, particularly when scales are added. Tick marks and scale labels improve the readability

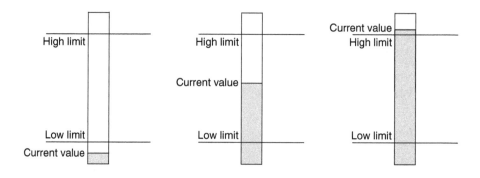

Figure 3.7 The bar graph.

of values from the graphic. The graphic has a distinctively different appearance in the too-high state and the too-low state. When reading too high, the graphic appears filled. When reading too low, the graphic appears empty. This is useful if users must distinguish between the states or take different actions. The two conditions, too high and too low, may differ in salience relative to other display objects, making one of the conditions more recognizable than the others. You may want to determine which condition is most critical and ensure that the background color of the bar will make this situation more salient.

Meter

Characteristics: The meter (Figure 3.8) shares many properties with the bar graph and is a development from traditional hard-wired indicators. Again, current value, limits, and distance to limits are easily shown. This graphic, however, does not have a significant color or shading change in its high or low states. This may be more appropriate in situations where the states are equally important.

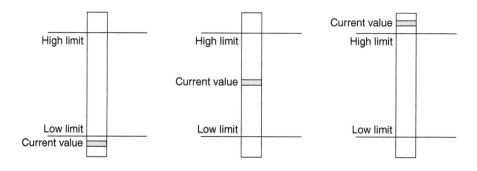

Figure 3.8 The meter.

4100.3	4523.5	4600.1

Figure 3.9   Digital display.

Digital Display

Characteristics: The digital display (Figure 3.9) is very compact. Precise values can be directly read from it, making it a good choice when users must record values or maintain a precise target. Values too high or too low can be shown using a secondary form of coding. Most often, color is used for coding, but size, font, or blink rates are also sometimes used. Showing distance to a limit would require either a second value or another form of coding.

Analog Plus Digital Display

Characteristics: This graphic (Figure 3.10) combines the advantages of the analog and digital forms. It is a good choice when operators must assess the general state of the variable (e.g., on target, too low, too high) but are also asked to make frequent reports or evaluations of the exact value of the variable. In situations where display space is a concern, this graphic may be divided with the digital information in a mouse-over or a pop-up window, or manually turned on and off. The analog form could be one of many possible forms.

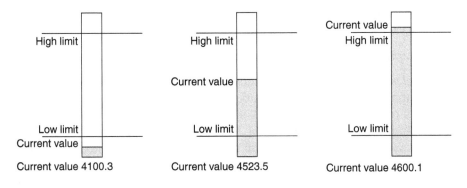

Figure 3.10   Combined analog and digital display.

Symbolic Display

Characteristics: In this display the variable is shown through a mapping to a different scale. In Figure 3.11, the scale used is color. This is in contrast to the general analog display that maps the variable to spatial distance. There are many different mappings that can be used (e.g.,

Figure 3.11   Symbolic display.

pattern, shading, size, blink rate, sound). The display can be very efficient in terms of space, and this can be combined with most other graphic forms. For example, a similar mapping was also used in the digital display example to indicate too-high or too-low regions. This kind of display is often used with alarms.

Variable with a Constraint

By a variable with a constraint, we are designing for all situations when something has a fixed limit that cannot be exceeded. Examples of this are items with a fixed capacity (like gas tanks and hard drives) and compositions of mixtures of two things that range from 0 to 100%. Generally users are interested in knowing the current value and also how close that value is to the constraint (e.g., "Is the tank nearly full?" "Are all the employee's hours assigned?").

The bar graph works well for constrained variables, where the top and bottom of the scale present constraints. Another analog form that works well in constrained situations, particularly where percentages are of interest, is the pie graph.

Pie Graph

Characteristics: The pie graph (Figure 3.12) is an analog form, just like the bar graph. In this case, the variable is mapped to polar coordinates, or degrees of arc, rather than to a linear scale. The graphic displays current value and distance to a constraint. Or alternatively the graphic can show two variables that are dependent on each other; e.g., amount full and amount available. Multiple variables can also be shown, provided though that variables are all related as partial measures of a larger variable. The nature of the graphic makes certain proportions easy to read (e.g., 25%,

Figure 3.12   The pie graph.

50%, 75%, 100%). More accurate readings of intermediate values may be difficult. High and low limits may be shown by shading or by limiting lines, but these may be more difficult to use than limiting lines on the bar graph.

Variable Where Normal Is Critical to Monitor

For high-level monitoring, often "normal" and "not normal" are key pieces of information. In this situation, we want to use a display that makes normal and abnormal conditions very salient. In most analog displays, such as the meter in Figure 3.13, we can indicate "normal" fairly easily. When operators are monitoring large processes with many variables, it can be useful to develop graphics that very quickly show the operator if the process is normal or not normal with very little other information. The purpose behind this strategy is to give critical information on as many variables as possible to construct an overview, then allow the operator to investigate off-normal variables in more detail on other displays.

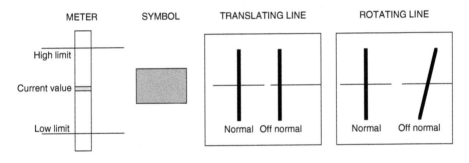

Figure 3.13   Indicating normality.

Meter

Characteristics: The meter (Figure 3.13, left) can allow normal and off-normal conditions to be determined. Generally, the normal should be positioned in the middle of the scale if the range permits that. If not, the variable can be normalized (divided by its normal value so normal = 1) or the normal value indicated graphically. Too-high and too-low information can also be shown and, depending on the complexity of the graphic, scale values given or a digital readout included.

Symbol

Characteristics: The symbol (Figure 3.13, center) works by mapping normal to another variable, in this case color. Thresholds for off-normal are determined and then mapped to other distinguishable colors, so that the symbol changes color when the variable crosses into the off-normal region.

Translating or Rotating Lines

Characteristics: These are effective displays where detailed information is not needed and normal and abnormal conditions are the main priority. The key to the graphic (Figure 3.13, right) is to map normal to a salient feature. In the translating line case, normal has been mapped to the midpoint of the line. In the rotating line case, normal has been mapped to the vertical or 90-degree position. The translating line depends on the perceptual ability to bisect the line to be interpreted. It is, in a very simple way, a type of meter, in this case horizontally positioned. It could just as easily be vertically positioned, in which case a horizontal line would move up and down, similar to the meter. The rotating line makes normal very salient by mapping it to the vertical or horizontal position. In this case, the mapping of off-normal high and off-normal low may be difficult to interpret. As well, a 180-degree change of position could be misinterpreted as normal.

Variable Changes with Time, or Rate of Change of Variable Is of Interest

Most variables change with time, and often the rate of change of variables is critical information. Something that is drifting slowly will be handled differently than a variable that is changing quickly. The rate of change may also be diagnostic and give clues to the source of a problem. Some cyclic variables may have regularly recurring patterns of values over time. We can show rates of change of a variable through trend charts and arrows.

Trend Charts

Characteristics: The trend chart (Figure 3.14) plots the value of a single variable over time. The current value of the variable can be obtained from the chart. The rate of change of the variable comes from the slope of the line. Whether the rate of change of the variable is constant (steady slope)

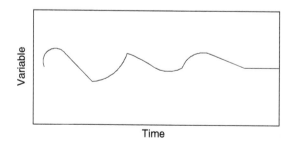

Figure 3.14   The trend chart.

or variable (changing slope) is also apparent. The chart can be enhanced with limit lines, scales, or background profiles to make normal values more salient. As well, more than one variable can be shown on the graphic.

Arrows

Characteristics: Arrows (Figure 3.15) can be used to add rate-of-change information to graphics that otherwise would not provide it. An up arrow shows increasing and a down arrow shows decreasing. If needed, the length or the width of the arrow can be used to show the magnitude of the rate of change, with longer or wider signifying a rapid increase. The pattern of the data cannot be extracted and changes in the rate of change may be difficult to interpret with this graphic.

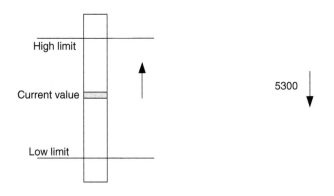

Figure 3.15   Indicating rate and direction of change.

Multivariate Display Options

Variable Balance, Variable = Variable

Many processes have variables that must be balanced or must remain equal to another value. Examples include flow measurements around closed loops, substances that must remain in certain proportions with each other, and expected movements of substances from one place to another. In showing these relationships, we want to show if the variables are equal, or if the proportions are changing. One way of showing this is through connected bar graphs. If time information is required, trend charts may be used.

Connected Bar Graphs

Characteristics: This is a modification of the basic bar graph to show the relationship between two variables. It is used most effectively when the variables are expected to be equal or in a consistent difference. The current value readings on the bar graphs are connected to each other and the slope of the line provides information on the relationship between the variables. The horizontal or vertical line of equal variables is most recognizable, and in some cases, the scales on the bar graphs may be modified to generate this feature. In cases, however, where it is important to monitor the difference between the variables and to retain knowledge of the current values, a sloped line may be used. Adding the background feature, seen on the two graphics on the right of Figure 3.16, makes deviations from this slope easier to follow. This graphic scales up well to multiple variable situations and can be used to create polar stars or configural displays.

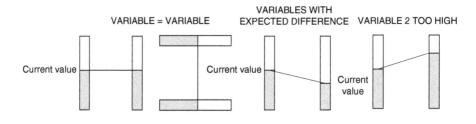

Figure 3.16   Showing balances with bar graphs.

Balances with Trend Charts

Characteristics: When showing balances on trend charts, the variables to be balanced can be shown on the same chart (Figure 3.17). This allows the individual values to be read and the difference between the values to be perceived by the distance between the lines. If the variables should remain in a constant difference from each other, or maintain at certain values, it may be useful to add a background profile to this graphic.

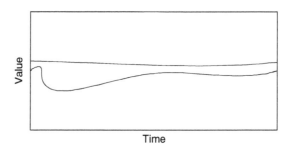

Figure 3.17   Showing balances with trend charts.

Variables Are Additive

When quantities sum, you often want to show the value of each variable and total value of the sum of the variables. Examples of situations with additive variables are resources of different types, flows into a tank from two sources, or multiple incoming lines of data. Choosing a design often depends on meeting additional criteria; for example, is it sufficient to show just the current values or does time information need to be shown? Two proposed alternatives are the summing bar graph and the summing trend chart.

Summing Bar Graph

Characteristics: The summing bar graph (Figure 3.18) stacks one value on top of another in a basic bar graph format. The individual variables can be identified by different shading, color, or patterns. Time history information is not shown. The total value or sum of the variables is readable from the top of the bar graph. The values of the individual variables are also available, although it can be difficult to read these values easily. In particular, reading variable 2 in Figure 3.18 would require subtracting the base value of the bar from the top value.

Figure 3.18   Showing additivity with a bar graph.

Nomographs

Characteristics: Nomographs (Figure 3.19) are the concept behind slide rules, but can be used today in graphical displays. Nomographs show various relationships using three scales and a linear line connecting the scales. Additive relationships are only one of many possibilities with nomographs; later we will show nomographs for multiplicative relations.

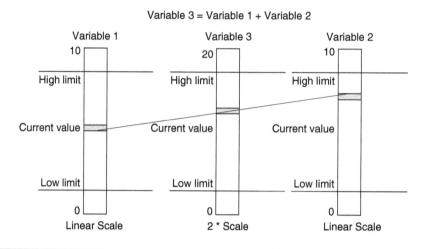

Variable 3 = Variable 1 + Variable 2

Figure 3.19 Showing additivity with nomographs.

Nomographs work by using different scaling techniques to develop the linear relationship. The strength of the display is in its strong linear feature developed by the connecting line. Variables 1 and 2 are easily read. To derive variable 3, however, a scale with twice the range must be used. Users may forget the change of scale with such scales in close proximity; therefore, reading exact values from the middle scale should be avoided.

Summing Trend Chart

Characteristics: The summing trend chart (Figure 3.20) is similar to the summing bar graph. The total value of the variables is readable from the highest level. Again, reading specific values on the individual variables can be difficult. One advantage of this form is that it shows the behavior of the variables and their totals over time.

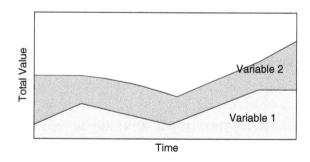

Figure 3.20 Showing additivity with trend charts.

Variables Are Multiplicative

Multiplicative relations take the form of a = b*c, or c = a/b. In these cases, you typically want to show all three variable values. Examples of multiplicative relations are frequent, like voltage (v = i*r) and heat transfer (q = mCpt). Graphical forms to show these relationships take advantage of geometric relations. Two options here are triangle graphics and nomographs.

Trigonometric or Triangle Relations

Characteristics: These graphics (Figure 3.21) make use of the trigonometric relationships of a triangle to map multiplicative relations. Two variables can be mapped directly and the third variable mapped to sin x, cos x, or tan x. When these relations have been used, the angular relation (e.g., tan x) is converted to a measure by using a line, fixed to the third side of the triangle that rotates as the angle of the triangle changes. In Figure 3.21, as x increases, the line attached to c maintains its attachment point and rises, thereby reflecting in the scale for variable 3. This graphic does not show time history information.

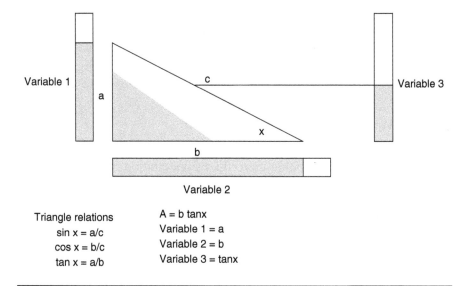

Triangle relations
sin x = a/c
cos x = b/c
tan x = a/b

A = b tanx
Variable 1 = a
Variable 2 = b
Variable 3 = tanx

Figure 3.21   Multiplication with trigonometric relations.

Nomographs

Characteristics: The strength of the display (Figure 3.22) is the ease of reading the values of variables 1 and 2, and the perceptible slope of the connecting line. The weakness of the display, in this case, is the logarithmic

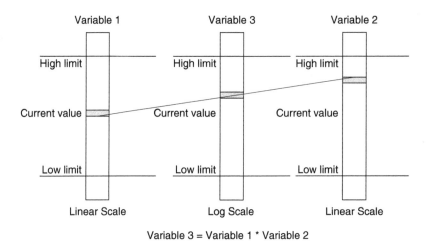

Figure 3.22   Multiplication with nomographs.

scale used for the product, variable 3. Logarithmic scales are challenging to construct and even more difficult to read accurately. Another weakness is that the product of the relationship, variable 3, is placed in between the other two variables, violating a typical reading pattern for the graphic (variable 3 = variable 1 * variable 2, or variable 1 * variable 2 = variable 3). This graphic should be used with care in situations where accurate reading of variable 3 is not critical or is supplemented with a second graphic.

Multiple Variables Determine System State

In many situations, users must integrate several variables to determine whether a process is operating correctly. Depending on the situation, they may need to read the individual values. In some situations, though, it may be sufficient to make a determination of system state, followed up with more specific graphics for diagnosis.

Configural Displays

Characteristics: These graphics (Figure 3.23, Figure 3.24) use multiple variables to create a recognizable shape. From variables plotted on a common background, the different values are connected by lines to form the configural shape. The shape may be a commonly familiar shape, such as in the polygon graphics, or it may be a shape that is meaningful to the particular process, as in the rankine cycle graphic. The purpose of these graphics is to provide a general determination of system state from many variables. The measures of each variable can still be seen, though

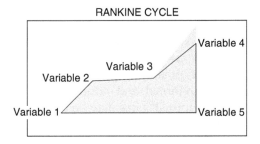

Figure 3.23   Configural graphics, the rankine cycle graph.

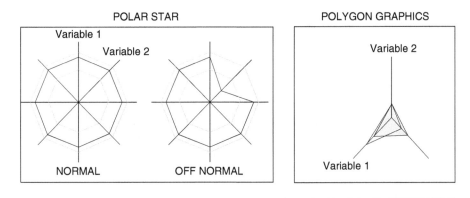

Figure 3.24   Polar graphics, the polar star and basic polygon graphics.

in most cases, the graphic should be supported by more specific information on individual variables. There are two graphical strategies to creating these shapes: The first is to connect the variables end to end (seen in Figure 3.23); the second is to connect the axes of measurement for the variables (seen in the polar star and polygon graphic in Figure 3.24). The polygon graphic may be easier to read individual values. Both graphics are enhanced by a background profile showing the shape of the process under normal conditions.

Polar Displays

Characteristics: Polar displays are a special case of configural graphics. They are created by orienting the axes of several variables so they originate from the same point. In this sense, they are an extension of the connected bar graph discussed earlier. Polar displays can be created with normalized data (the polar star) or not-normalized data (the polygon graphic). A particular feature of the polar star is that all the variables are normalized, and normal is set to the same distance on each axis. In this way, the polar star forms a regular polygon when the system state is normal and an irregular polygon in abnormal states. In contrast, with the polygon graphic, the axes are not normalized. In the polygon graphic, the normal condition of the polygon could be an irregular shape, which may be difficult to identify. In the polar star, the regularity of the shape makes the difference between abnormal and normal easily identified. Furthermore, specific types of abnormal states may be recognized from the shape of the polar star. The polar star should be tested to ensure that various anticipated problem states generate distinctive irregular polygons. In many cases this is the advantage of the polar star over the basic polygon graphic.

The polygon graphic, however, can be used very effectively in situations where process values are typically low and any high value indicates a potential problem. In this situation, the polygon graphic develops an emergent point towards the variable of interest. Specific values are difficult to read from the polar star graphic (though possibly easier with the polygon graphic). The polar star should be supplemented with more detailed displays. In both cases time history information is difficult to add, although sometimes arrows have been added to show the direction and rate of change. Both graphics in Figure 3.24 are enhanced by showing high and low limit values with either either lines (shown in the polar star) or a background profile, shown in the polygon.

Bar Graphs with Configural Features

Characteristics: Bar graphs of multiple variables (Figure 3.25) can be aligned to show patterns in the data. The expected change of values from high to low can create a pattern that users can recognize over time. These

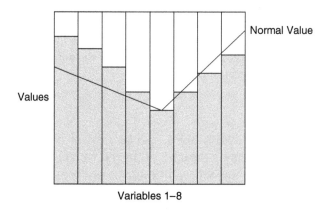

Figure 3.25   Configural features of a bar graph.

normal values can be enhanced by providing a reference for normal values (the line). For this kind of graphic to be successful, the variables have to be of the same type and have similar ranges so they can be placed on a common scale. Time history data is not shown.

Line Graphs with Configural Features

Characteristics: The characteristics of this graphic (Figure 3.26) are similar to the configural bar graphs. In this case lines are used instead of filled bars. The concept remains similar, though; components are placed next to each other and values plotted on the graph. All components need to have similar measures. This graphic is most effective for long lines of relatively similar components, where the net output from the line of components is of interest.

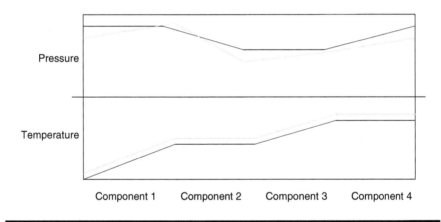

Figure 3.26 Configural features of a line graph.

Mass Data Display

Characteristics: The mass data display is a very efficient way to display many variables. Each variable is described by a simple symbol (line, circle, square, and so on) that changes properties as the variable changes. Properties that may change are angle (shown in Figure 3.27), size, color, thickness, and so on. The purpose of the graphic is to organize the symbols effectively so that changes in the data are quickly recognized. The most effective mass data displays will also generate recognizable patterns for anticipated faults. While specific values cannot be read from the diagram, overall system state is quickly recognizable, and problem areas can be identified. This is a very effective display for high-level monitoring of complex processes.

INDIVIDUAL HORIZONTAL LINES   JOINED VERTICAL LINES   SYMBOLIC TILES

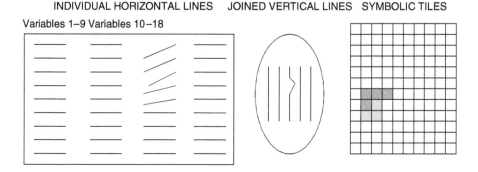

Figure 3.27   Mass data display.

Multiple Variable Balance

In showing a balance across several variables, the graphic needs to show the individual variables as well as whether or not they are in balance. For the high-level monitoring situation, showing the balance is more important. Two ways of doing this are by using meters in close proximity and by using meters connected with lines.

Close Proximity Meters

Characteristics: If several variables must balance to the same value, putting bar graphs in tight proximity to each other is a possible approach (Figure 3.28). The graphics form an emergent feature, that of a horizontal line, when the variables are balancing to the same value. Depending on the size of the graphic and the scale used, users may be able to read the values from the graphic. Lines indicating normal values or high or low thresholds will improve the graphic.

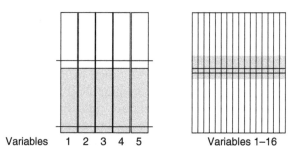

Figure 3.28   Close proximity meters.

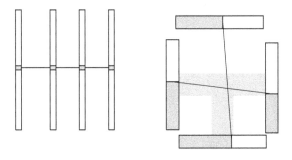

Figure 3.29   Connected meters.

Connected Meters

Characteristics: In these graphics (Figure 3.29, left), several meters are connected together by lines drawn to their current values. The lines form an emergent feature that should indicate if the values are balanced. Typically, horizontal or vertical lines are used since these form easily recognized features. Other constructions may also be useful, such as the cross-hair balance (Figure 3.29, right) designed by Dinadis (2002). Size and scaling will affect the readability of the individual variables from the graphic. The graphic can be improved by indicating normal values and high and low thresholds.

Multiple Variables with Interacting Constraints

In some cases where multiple variables have constraints, the variables are related to each other. When the variables are related, it makes sense to display them against the background of the constraints. This background gives meaning to the data and shows how close the variables are to their constraining limits. These constraints can often be derived from the physical laws governing the process you are displaying.

Variables on a Background of Constraints

Characteristics: In graphics of this type, typically two or more measures are needed to determine the state of the process at that point. In particular, what state the process will be in depends on the physical laws governing the process. For example, in Figure 3.30, right, the measurements of temperature, pressure, and entropy determine whether water is liquid, vapor, or a mixture. The background for the graphic is the physical properties of water. Similarly, in the left graphic, the rate of production is determined by process temperature and feed rate. There are constraints

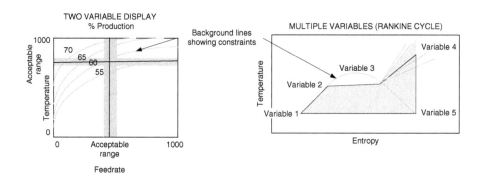

Figure 3.30   Variables against a background.

limiting what temperatures and feed rates are possible, shown by the shaded regions. These physical constraints therefore limit the possible production rate. The key to these graphics is to determine the physical properties of the process and to use these properties as a background for plotting plant values.

Structural Display Options

Linear Structures

Linear structures need to portray sequencing and order. They are needed when showing flow between components, or variables that fall in a certain order. They have a clear starting point and a clear ending point. Users need a representation that they can trace from start to end, and identify changes over the sequence or changes that cascade from one part to another. Most common graphs, bar charts, and trend charts, including those that we have already seen, employ linear representations on their various axes. In this sense, we are talking more of ordering variables or developing a mapping across a complex display. Two ways of doing this are through lines and spirals.

Lines

Characteristics: Lines include any linear ordering (Figure 3.31). This includes determining what elements go on the left of the display page and what elements go on the right. Linear orderings can also include vertical structures. They may be connected by arrows, emphasized by axes if the linearity is quantifiable or categorizable, or just inherent in the

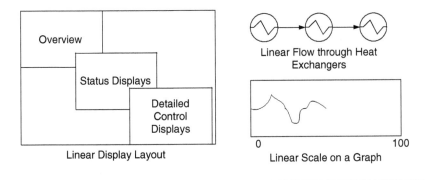

Figure 3.31   Linear arrangements.

layout of the display. Timelines are an example of linear ordering. In general, there are a few rules:

- Flows of material, information, and sequences of components should go left to right, and up to down.
- Overview displays should be located at the top or top left of the display, with more detailed information to the lower right.
- Lower values and earlier events are generally located to the left.

Spirals

Characteristics: Spirals (Figure 3.32) are a form of linear ordering in a circular form. Spirals differ from circles by having a start point and an end-point (this is why they can be considered linear) and a certain amount of outward translation every 360 degrees. Spirals can be used to compress large amounts of linear data. Spirals can also be used when data has some cyclicity to it (that is, when a certain pattern repeats itself).

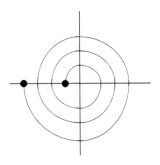

Figure 3.32   A spiral.

Spatial Structures

Spatial structures often benefit from using 2D or 3D space as their basis of organization. Parameters are mapped onto x and y coordinates either quantitatively or in a looser categorical mapping. Examples of spatial representations are maps and matrices.

Maps

Characteristics: Maps (Figure 3.33) are used to show the layout of components in general relation to their actual position in the world. Maps are a good approach when users must leave their interface to examine components in the world or must communicate with workers in the field. Maps are also a good visual approach in all cases where material changes location. Maps can be presented at various levels of detail, with less-detailed maps serving as overview displays and navigation tools and more-detailed maps allowing actual control over components.

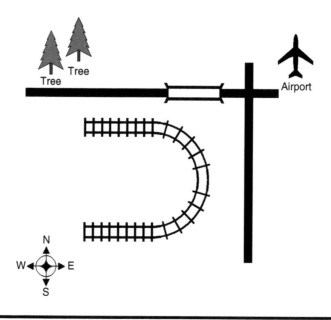

Figure 3.33   Maps.

Matrices

Characteristics: Matrices are 2D orderings (Figure 3.34). The orderings may reflect location in the world or any variety of other variables. Matrices are useful for showing structures that are described by two variables that are

Height

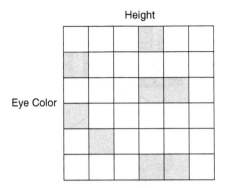

Eye Color

Figure 3.34   Matrix with independent dimensions.

independent, where there is an element present for most of the cells described by the two variables.

Symbolic Structures

Symbolic structures have an underlying set of abstract relationships. The nature of the relationship could be nearly anything. The issue is showing the relationship between elements and allowing a user to trace through the relationship. The forms we discuss, trees and networks, are differentiated by the nature of the structures that they represent. We have included mimics and other diagrams as another example of representations of symbolic structures.

Trees

Characteristics: Trees (Figure 3.35) are useful for portraying hierarchies. The structures need to have a start point and an end-point that are mapped to the top and bottom of the tree. There must be consistent levels that can be differentiated from one another. There is a reason for stepping from level to level, and these steps generally do not skip levels or move backwards. The Abstraction Hierarchy has been shown with a tree in many cases in this book. Trees are also used for organizational charts and tables of contents.

Networks

Characteristics: Unlike trees, networks (Figure 3.36) have many different interconnections. Levels are not as rigidly defined, although there may be looser structures, such as subnetworks. In many cases, networks are used

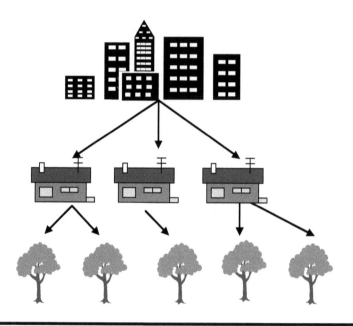

Figure 3.35   Hierarchy of elements in a tree representation.

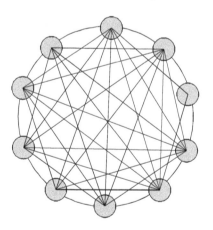

Figure 3.36   A network arrangement.

to portray situations with many connections, such as data networks or social networks.

Mimics and Other Diagrams

Characteristics: Mimic diagrams (Figure 3.37) are a special kind of symbolic map. Mimic diagrams use symbols to represent equipment and

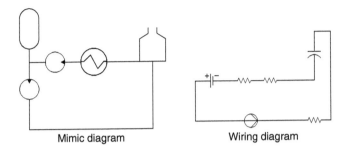

Mimic diagram   Wiring diagram

Figure 3.37   Mimic diagrams and other diagrams.

portray the connections. The layout is usually spatial, but this may reflect the connection of the equipment and not particularly the location of the equipment in the world.

HOW TO DEVELOP A NEW VISUAL FORM

The visual forms in this chapter are only examples, and many more visual forms exist. We can't tell you exactly how to create a new visual form but we can offer some general directions.

Do you need to show a single value or multiple values together? Generally if people need to relate multiple values to each other, you'll want your visual form to do this. Remembering many values and integrating them creates a lot of mental workload, so the goal is to design a display that can do this work. You need to decide whether people need to read the individual values exactly or just have an overall sense of their state.

Are there any physical principles or textbook graphs that show how the values relate? Many excellent EID graphics have been implementations of graphs, or plots, or charts from physics textbooks, biology textbooks, medical textbooks, and so on. Don't overlook these sources of inspiration. Great examples of relating information visually are available (e.g., Tufte 1983; Tufte 1990; Tufte 1997).

Are there any geometric relationships that you can use? In many cases, we have taken advantage of geometry. Separate constants from variables, and examine the basics of the equation you are planning to represent. Match the equation to a geometric shape, and you may be able to develop an object display that shows properties that are useful. Figure 3.38 contains a summary of geometric relationships.

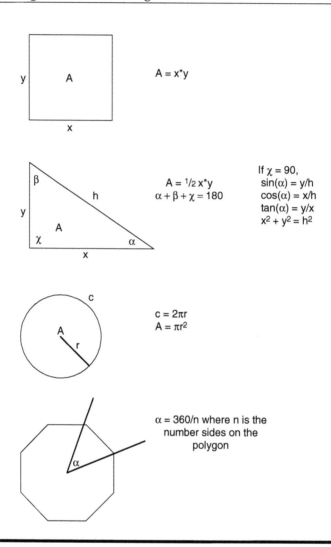

Figure 3.38   Common geometric relationships.

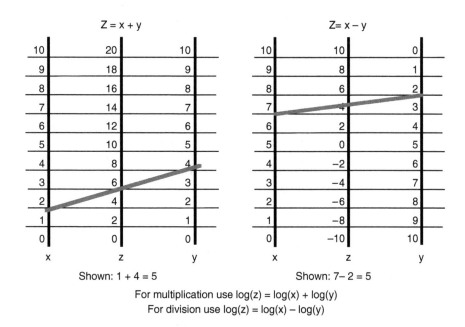

Figure 3.39   Basic nomographs.

SUMMARY: VISUAL THESAURUS REVIEW SHEET

Single Variable Display Options

Display Need	Visual Options	Best Use	Major Limitation
Variable within limits	Bar graph	Showing context	Precise reading
	Meter	Showing context	Precise reading
	Digital display	Precise reading	Context
	Analog plus digital display	Precision and context	Space
	Symbolic display	Tight spaces	Learning meaning
Variable with a constraint	Pie graph	Proportions, fractions	Precise reading, quantity
	Bar graph	Distance to limit	Precise reading
Variable where normal is critical to monitor	Meter	Distance to limits	Normalizing the scale
	Symbol	Alarms	Precise meaning
	Translating line	Fault detection	Precise reading
	Rotating line	Fault detection	180-degree problem
Variable changes with time or rate of change is of interest	Trend chart	Historical info, rate of change assessment	Space requirements
	Arrows	Tight spaces	No data pattern displayed

Multivariate Display Options

Display Need	Visual Options	Best Use	Major Limitation
Variable balance, variable = variable	Connected bar graphs	Showing deviation from balance	Precise reading
	Trend charts	Balance over time	Precise deviation
Variables are additive x = y + z	Summing bar graph	Reading x	Reading component values (y, z)
	Nomograph	Calculating new values (x, y, z)	Space
	Summing trend chart	Summing over time	Space and readability
Variables are multiplicative x = y * z	Triangle graphics	Showing complex relations	No time history, training
	Nomographs	Component (y, z) readability	Reading of log scale (x)
Multiple variables determine system state	Configural displays	System state	Time history, reading individual values
	Polar displays	Normal/off-normal	Extracting values, time history
	Bar graphs with configural features	Sequential similar readings	Precise reading
	Line graphs with configural features	Sequential similar readings	Precise reading
	Mass data diagrams	Large amounts of data	Time history, precise reading
Multiple variable balance	Close proximity meters	Same range variables	Time history
	Connected meters	Showing deviation	Time history
Multiple variables with interacting constraints	Variables on a background of constraints	Showing context	Time history

Structural Display Options

Display Need	Visual Options	Best Use
Linear structure	Lines	Order or flow
	Spirals	Large ordered data sets or ordered data sets with underlying periodicity
Spatial structure	Maps	Navigation, overviews, showing location
	Matrices	2D structures
Symbolic structure	Trees	Hierarchies
	Networks	Networks, large data sets
	Mimics and other diagrams	Showing component connections

4

USING A WORK DOMAIN
MODEL IN DESIGN

Now that we have a basic language for describing our designs, how do we take the information from our Abstraction Hierarchy and turn it into effective display graphics? The Abstraction Hierarchy has given us a five-level description of the work domain. From the analysis we have learned the bounds of the work domain (what to include and not include), viewed the work domain from five different perspectives, and connected those five levels with means-end links. From the analysis we can extract information requirements, key constraining relationships, multivariate relationships, and means-end relationships. These are the basic information elements for the design of our graphical displays.

INFORMATION REQUIREMENTS

Extracting information requirements from an Abstraction Hierarchy involves converting the model into a list of variables (Figure 4.1). Level by level, each part of the work domain model represents a specific variable set (i.e., Physical Form variables, Physical Function variables, process variables, and Abstract Function variables). Table 4.1 shows some typical variables at each of the four levels.

You will notice that Functional Purpose variables are not described in that table. Functional Purpose variables are almost always represented in one of the other four levels. For example, it may be a purpose of a system to reach a certain temperature, generate a certain force, or keep something at a certain level.

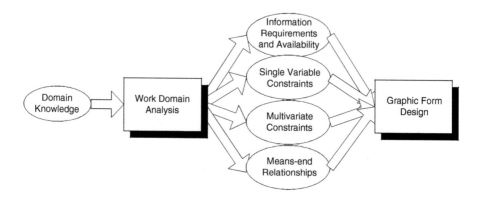

Figure 4.1   Systematic approach to graphic form design.

Table 4.1   Variables at Different Levels of the Work Domain Model

Physical Form	Physical Function	Generalized Function	Abstract Function
Color	Level	Temperature	Mass
Shape	% open or closed	Pressure	Energy
Size, length, width	Distance	Volume	Momentum
Depth	Server capacity	Velocity	Force
Server location		Acceleration	Power
		Software process measures	Torque
			Information

Let us return to the example of the car from Chapter 2 (shown in Figure 4.2). In extracting information requirements from the model, we work through the model systematically, extracting variables level by level.

Table 4.2 is a partial list of variables that can be extracted from the model. It is not a complete list.

In generating the list, we worked through each level of the hierarchy. For example, given the box of "Conservations of Mass and Mass Flow," we derived the variables Mass of car, Mass of people, Mass of cargo, and Rate of mass transfer. This could have been broken down to further detail, working out the mass of all parts of the car, the mass of air intake, the mass of exhaust out, and so on. When we design our interface, these are the variables that could be included, depending on the purpose and level of detail of the display. From Gas tank, at Physical Form, we derived Gas tank level and Gas tank percentage full.

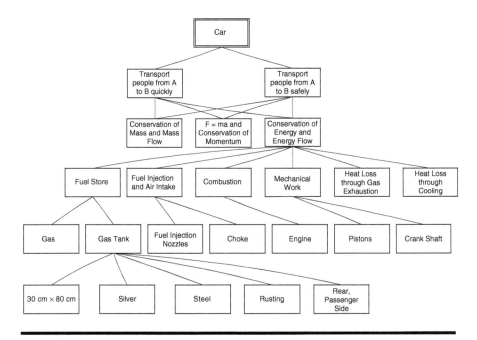

Figure 4.2   The car example from Chapter 2.

The method for generating the variable list is to work methodically through each box in the hierarchy. At each box, we ask, "How could we measure that?" We are trying to learn specific information about the element. Knowing the level that we are examining allows us to focus that inquiry to generate a thorough list of requirements. In the next sections, we move through level by level, providing the types of questions to ask and measures to collect.

Types of Measures at Each Level

Functional Purpose

At this level, we are looking for overall measures of system performance. How can we tell that the system is working as designed? In the car example, we identified two related purposes: "Transport people from A to B quickly" and "Transport people from A to B safely." How can we measure these purposes? We can measure how many people we have transported and how long it took to transport them. We could measure what distance they were transported, or how close they got to location B. The measurement of safety, though, can be more challenging. Accident or injury rates may be measures here.

Table 4.2   Variables at Different Levels of the Model, Car Example

Level	Variables
Functional Purpose	# of people transported
	Time to transport
Abstract Function	Mass of car
	Mass of people
	Mass of cargo
	Rate of mass transfer
	Force generated
	Torque
	Energy in
	Energy out (kinetic)
	Energy out (waste: heat and exhaust)
	Power (rate of energy use)
Generalized Function	Volume of fuel stored
	Fuel injection rate
	Air flow rate
	Engine pressure
	Engine temperature
	Exhaust temperature
	Piston displacement speed
	Speed of car
	Distance traveled by car
	Time to transport
Physical Function	Gas tank level
	Gas tank percentage full
	Choke percentage open
	Piston displacement
	Passengers
Physical Form	Size of gas tank
	Size of car
	Number of passengers

Abstract Function

At this level we are always concerned with flows, balances, and conservation. So it makes sense to collect measures on all inputs, storage levels, and outputs in terms of mass, energy, or information. We also need to collect transfer rates for mass and energy. We need to know what mass and energy elements flow and to where they flow.

Generalized Function

Generalized Function is usually measured through process variables. In physical processes, temperatures, pressures, and volumes are useful measures. Speed and distance are useful in transportation systems.

Physical Function

At Physical Function we are looking for measures of capacity and capability. Tanks, for example, have levels, and a percentage of capacity (full, empty, or partly full). One way to find these variables is to look at the possible equipment settings and possible states. Can it be on or off? Open, closed, or partly open?

Physical Form

At Physical Form we list all the physical attributes that could be measured about the equipment. The list here can be extensive: size, color, shape, material, condition, location in x, y, z. For example, the car may have certain dimensions and shape, it may be red, and it may be made of steel.

INFORMATION AVAILABILITY ANALYSIS: WHAT TO DO IF YOU CAN'T MEASURE IT

It is critically important when generating information requirements from your work domain model that you do so without concern for whether or not you can physically measure the variable. This may seem counterintuitive, or perhaps even a waste of time, but it is a very important aspect of ecological design. Ecological design must always be based on how the environment actually works. Therefore, if something changes temperature, but there is no temperature sensor, you should still record that temperature in your information requirements. If something is hard to measure, like safety, put in a placeholder like "Measurements of safety" until you can refine the measures later. The objective is to keep your list of requirements as broad as possible at this stage in your design.

There are several very important reasons for keeping this breadth in your analysis. While you may not be able to measure the variables now, in the future additional sensors may be available that can make those measurements. You may be able to determine a way to calculate the measure from other variables. While you will obviously have to design an interface using the information that is available, your broader list of information requirements provides a resource for the future, if the opportunity came to upgrade your system.

An information availability analysis is a structured way to look at the information asked for in your work domain model. It is important to identify what is currently sensed, what could be calculated, and what variables must be left for developments of the system (Dinadis and Vicente 1999). In the past (Moradi-Nadimian et al. 2002) we have examined:

1. Variables currently measured by sensors
2. Variables calculated from sensor data
3. Variables not currently measured
4. Variables that could be calculated

If there are different kinds of instrumentation (e.g., an on-board diagnostic system vs. an off-board analysis system, real-time vs. batch systems), it may be useful to differentiate these sources as well.

If we go back to the previous variable list, we differentiate between the variables along these categories. In Table 4.3 we used **bold** to indicate variables that are not sensed, *italic* to indicate variables that are not currently available but could be calculated, and normal typeface to indicate sensed variables. We assumed the car had an on-board diagnostic system for engine performance.

If we were interpreting this list we would see that our hierarchy has identified many variables that we may not be able to use directly. Some of these we may be able to calculate or may be available in future cars. For example, weight sensors are beginning to be developed to control the deployment of air bags. These could also be used to monitor the number of passengers and mass of passengers and cargo. With a little extra effort, we may be able to calculate some useful measures. For example, energy usage and energy waste are variables that may be more useful in the future, because of rising energy costs.

At this stage, you would normally go back to your engineers and see if any of these sensors or calculations could be added. Then you may want to work towards two sets of interface drawings — one using the current information set, and one showing the advantages of an improved information set.

USING INFORMATION REQUIREMENTS IN DESIGN

If we review the visual forms that we discussed in Chapter 3, most of your information requirements can be displayed using single variable graphics. Bars, meters, and symbols will work in most cases. This is the simplest way to begin your design.

As you work through the next passes, extracting single variable constraints and multivariable constraints, you will gain the information needed

Table 4.3   Information Availability, Car Example

Level	Variables
Functional purpose	**# of people transported**
	Time to transport
Abstract function	Mass of car
	Mass of people
	Mass of cargo
	Rate of mass transfer
	Force generated
	Torque
	Energy in
	Energy out (kinetic)
	Energy out (waste: heat and exhaust)
	Power (rate of energy use)
Generalized function	Volume of fuel stored
	Fuel injection rate
	Air flow rate
	Engine pressure
	Engine temperature
	Exhaust temperature
	Piston displacement speed
	Speed of car
	Distance traveled by car
	Time to transport
Physical function	Gas tank level
	Gas tank percentage full
	Choke percentage open
	Piston displacement
	Passengers
Physical form	Size of gas tank
	Size of car
	Number of passengers

to develop more complex graphic forms. Single variable constraints are used to develop context and limits for your graphics. Multivariable constraints can be used to develop more complex graphic forms, like configural graphics.

Single Variable Constraints

Display variables are only one level of information you can collect from your model. Your model can also be looked at in terms of constraints.

Constraints add another dimension to information extraction. When we extract constraints we begin to understand how that system works and what are its limitations. We learn its maximum performance level and minimum performance level. Most importantly, we learn what happens when the system is broken. To extract constraints we take a second pass through our model, asking, "How much?," "How quickly?," and "Is there a limit?"

Questions to Elicit Single Variable Constraints

Working through the car example again, the following are the kinds of questions we would ask.

Functional Purpose

How many people? How quickly? How far? How safely?

Abstract Function

Maximum or minimum mass limits? Maximum or minimum energy limits? Limits on acceleration or momentum?

Generalized Function

Maximum or minimum fuel volume? How much air intake? Constraints on rate of fuel use? Maximum or minimum engine pressure? Maximum or minimum engine temperature? Maximum speed of car?

Physical Function

Maximum or minimum gas tank level? Maximum or minimum choke positions? Maximum or minimum piston displacement?

Physical Form

Maximum or minimum number of passengers, size of car, and so on.

From these questions, we build a table of constraining relationships at each level, as in Table 4.4. Again, these are examples of constraints for this work domain, but clearly not a complete set.

Using Single Variable Constraints in Design

We collect the constraint information because it is very useful in designing our interface. These constraints bring a lot of meaning to the requirements

Table 4.4   Single Variable Constraints at each Level, Car Example

Level	Constraints
Functional purpose	1 < # of people < 6 Max distance (per tank of gas) = 400 km 50 km/hr < speed < 140 km/hr
Abstract function	50 kg < Mass of car contents and tow < 1000 kg Power < 145 hp Acceleration < 0 to 100 km/hr in 10 s
Generalized function	0 L < Fuel volume < 50 L 5 m³/s < Air flow rate < 10 m³/s Fuel use < .1L/km Engine temperature < 500 deg C Engine pressure < 300 kPa Max speed of car < 140 km/hr
Physical function	10% < Gas tank level < 100% 0% < Choke < 100% Max piston displacement < 3 cm
Physical form	Safety belts = 5 Length of car = 2 m

we specified earlier. Now not only do we know what measures to use, we also know what levels are high and low, good and bad. These are the foundations of establishing context.

We use these constraints to generate backgrounds for reading our data. They determine what the ranges of our scales are, where alarm limits are, and what the important thresholds are. From these constraints we can design background lines and profiles, and determine different visual coding schemes.

Multivariate Constraints

Our next pass through our hierarchy looks for multivariate constraints. By this, we mean relationships of two or more variables. These relationships may occur at the same abstraction level, or they may cross levels.

Multivariate relationships are often expressed in the form of equations. These are the relationships you find in textbooks and learn about in school. For example, conservation equations always relate at least three variables: an input, an output, and a storage (In - Out = Stored). Three variables are immediately defined by this equation. The equation is a constraint, because the laws of physics mandate that the variables are related in this way. The equation holds true for anything that does not change form, be it mass, energy, money, or cookies. Since all the elements of the equation

Table 4.5   Examples of Multivariate Constraints by Level

Equation	AF	GF	PFn	PFo
Heat transfer $Q = m*Cp*dT$	m, Q	dT	Cp	
Momentum mu = mv	Mu, m	v		
Force F = ma	F, m	a	Material	
Torque = F*d	F	d		
Ideal Gas Law pV = nRT		p, V, T	R	n
Nuclear Decay $N = N_o e^{-(0.693t/T_f)}$	N, No	t	T_f	
$E = mc^2$		E, m	C, medium	

are at the Abstract Function level, this is a multivariate relationship at a single level of abstraction.

Determining the physical equations behind a process is the easiest way to obtain multivariate constraints. In some cases, you may find that the relationship is so complex, there are no equations available to describe it yet. Some of these relationships have been determined empirically. By this we mean that people have studied how variable X and Y create variable Z by experimenting and recording the various Zs for the combinations of Xs and Ys. The determination of the state of water from temperature, pressure, and entropy is an example of this. In these situations, there are usually charts or tables available for looking up this information.

Most interesting perhaps are relationships that express means-end links or relationships that cross abstraction levels. These relationships are found when an equation uses two variables at different levels of your hierarchy. The interesting thing about these equations is that they express how lower level elements work towards higher level purposes.

Table 4.5 shows examples of some of these constraints. We have also isolated the elements by abstraction level.

From our car example, we can derive several multivariate constraints. Stated generally, some of the relationships are:

Combustion efficiency = f(fuel flow, oxygen flow, engine temperature, and pressure)
Acceleration = f(torque) = f(engine pressure)
Torque = f(piston displacement, piston speed) = f(engine pressure)
Mass (people getting in car) = Mass (people getting out of car) = Mass (people in car)

Using Multivariate Constraints in Design

Multivariate constraints are the basis for some of the more interesting configural displays. Well-understood relationships can take advantage of additive or multiplicative graphic forms. Relationships that are derived from empirical data can use that background of empirical results to add context to their display. Relationships that are present, but for which equations and empirical results are not available, may benefit from a graphic like the polar star, which indicates that the axial variables are all used in determining some more global state.

Visually showing multivariate constraints is possibly the strongest aspect of ecological design. These relationships are often complex and hard to understand. They may require calculation or looking things up in tables or manuals to confirm. Showing these relationships on your interfaces reduces mental workload dramatically and may also help your users to understand the system that they are monitoring even better.

Means-End Relationships

The final type of information you gain from your hierarchy is means-end relationships. These come from the lines you have joining elements across levels. A means-end relationship suggests that one element is implicated in the value of another element. We discussed some of these relationships in the previous section as multivariate constraints.

You may not be able to find equations for all your means-end relationships. You will still want to show these relationships, though, since this is how your user can work towards achieving system purpose or diagnose faults.

If we work through the car example, we can make a list of means-end relationships from our work domain model. Table 4.6 gives a chain of means-end relationships.

This chain of relationships gives information important to your display. In order to know how far they can transport people, and if they can even

Table 4.6   Means-End Relationships, Car Example

Means	End
Conservation of energy and energy flow	Transport people from A to B quickly
Fuel store, injection, combustion	Conservation of energy and energy flow
Gas, gas tank	Fuel store
Size, color, material	Gas tank

reach B, your users need to know how much energy they have. The amount of energy is determined by the amount of fuel. The amount of fuel is stored in the gas tank, which is full to a certain level. The gas tank has a certain capacity.

Using Means-End Relationships in Design

Knowing your means-end relationships helps you to search for key multivariate constraints. Even more generally, though, means-end relationships help you organize your display and group relevant graphics together. Means-end relationships also allow you to determine salience levels across your graphic forms.

EXAMPLE DESIGN PROCESS

In developing an ecological design, it is often useful to work through the previous stages graphically. Start with defining your information requirements and then work towards more complex graphical objects. The advantages of working in this manner are:

1. You begin your design as early as possible.
2. You minimize the chances of getting mental blocks while designing, since there is always a place to start.
3. Leaving more complex objects to the end takes advantage of any additional relationships you have learned while working with the system. (Be sure to update your work domain model, to keep a record.)

Stage 1: Basic Design of Information Requirements

The simplest way to complete this stage is to convert every information requirement into a basic graphic. Often, meters and bar graphs are a reasonable way to start.

Continuing to work with the car example above, we will focus on the fuel system (Figure 4.3).

We have taken a sample of the Abstraction Hierarchy variables at four levels to use in our example. At this stage, we have just shown each of them as meters, using bar graphs, or digitally. Obviously this won't be the final display, but it's a place to start. We have already been able to simplify our information: Gas tank level and gas tank percentage full can both be shown using the bar graph. Where the height of the bar shows the level, the percentage of gray maps to the percentage full. For organization we have laid out the meters following the general flow; gas tank

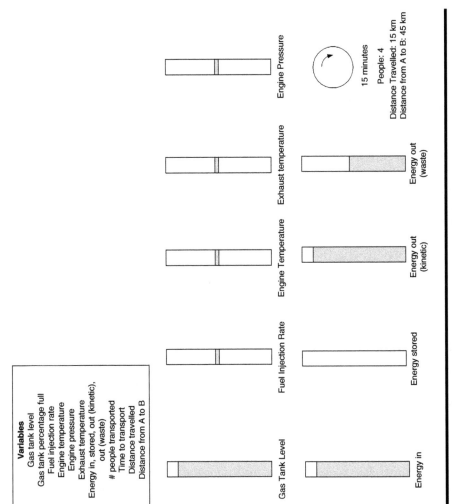

Figure 4.3   Basic information requirements to graphics, car example.

to exhaust flows left to right. In the second row we have shown higher level information. We chose a circular meter for time to map to a clock with a digital readout, presuming people want high accuracy in estimating time. We also chose a digital readout for the number of people being transported: the distance traveled and the distance between A and B. We have, for the sake of the example, assumed we can get measures of all this information. This is a Stage 1 design, showing information requirements.

Stage 2: Single Variable Constraints

When we add single variable constraints we are looking at adding information that enriches our basic variables. We are adding information on how high, how low, too high, and too low. In the draft display (Figure 4.4) we have done this by adding threshold lines for alarm limits and shading to show normal regions. We changed the display of the number of people to a block of six symbols that change shading. In contrast to the previous display, this one shows more clearly that the car is carrying four out of a possible six people. The graphic is different than, but compatible with, the bar graphs already used. We also realized that distance traveled and distance from A to B could be expressed as a single variable. We used a horizontal bar graph showing the percentage of distance traveled. We used the distance from A to B as a constraint. We maintained the digital display, since this is something where people may want exact values.

Stage 3: Multivariate Constraints

When designing in multivariate constraints, we look for relationships between variables. What variables balance, and which variables work together to determine the state of a system or a process? We also look at which variables need to be viewed across time.

In the display in Figure 4.5, we have identified a few possible multivariate relationships and modified our design further. Engine pressure and temperature relate to each other and determine the state of the engine, so these variables should be displayed together. Our energy in and out measures reflect an energy balance, so these variables have been linked with a line between the graphs that is horizontal when the values are properly balanced. We had to normalize the bar graphs in order to create this horizontal feature. The gray shading behind the line shows the expected normal values. We removed energy stored, since under normal circumstances, energy stored is negligible and an imbalance in energy will show on our balance graphic. From our higher level information, we

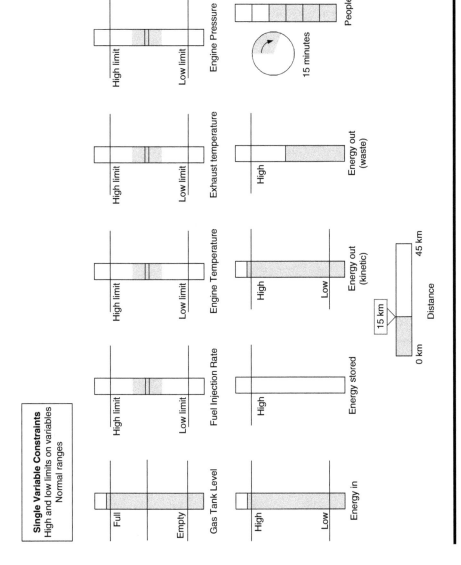

Figure 4.4   Single variable constraints to graphics, car example.

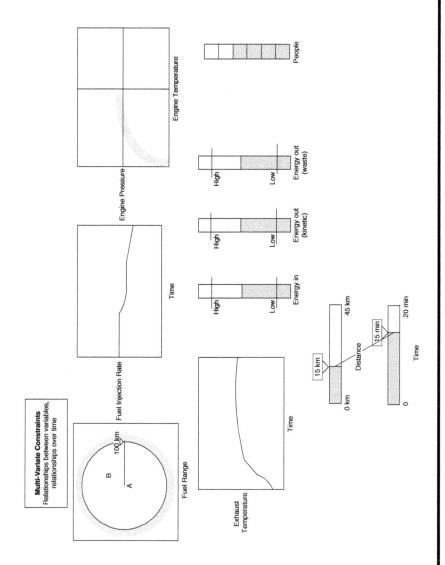

Figure 4.5   Multivariate constraints to graphics, car example.

realized that distance and time are related, through speed. We modified our time graphic to be more compatible with the distance graphic. Now we can balance the two graphics, such that travel at the expected speed will generate a vertical line. Already something is apparent from the values we have been using in our graphic: We are running late!

We also related fuel amount and distance traveled to show whether we had enough fuel to reach B. To improve the graphic, and to keep the limits we used before, we added a shaded circle showing the maximum travel possible on a full tank of gas. As the car travels further, the white inner circle will shrink, making the lowered gas level more salient.

In trending information, we provided two trends: one for fuel injection rate (to show the possible clogging of fuel injection nozzles), and the other for exhaust temperature (to show the vehicle warming up).

Stage 4: Organize by Means-End Links

In the final stage of our design we review our work domain model again. This time we are looking for clues on how to organize our display, how to connect our graphics, and how to focus our user's attention. In general, we want our users to monitor at the highest level of our work domain model, at Functional Purpose. Under normal conditions, monitoring at this level is adequate, and users only need to move to other levels if problems arise. Therefore, we take the graphics that relate to this level and place them in the top left corner of our screen so they will be read first. We make these graphics larger and more salient. We put less important graphics below or to the right.

In our example (Figure 4.6), the graphics that relate to our purpose are the distance/time graphic and the fuel range graphic — these let us know that the car is going to get the people from A to B and on time. We placed these at the top of our screen and grouped them with a background shading. We kept strong shades of gray on these graphics to improve their salience. We reduced the size of the graphic for the number of people. This is most often a constant and does not need regular monitoring.

Our other graphics support the Abstract Function and Generalized Function level. We organized these top to bottom, by Abstraction Hierarchy, and left to right by process flow (i.e., from fuel to engine to exhaust). We emphasized the means-end connection between these graphics by shading in between the levels, so that the user can clearly see which Generalized Function graphic relates to which Abstract Function graphic. We decreased the salience of all these graphics by reducing their size and lightening up the gray scale. Under normal conditions, our users won't need to monitor these graphics. The information is available, though, in

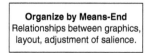

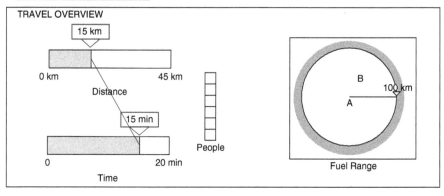

PROCESS DISPLAYS

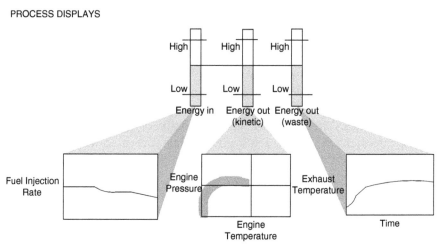

Figure 4.6 **Means-end relationships to graphics, car example.**

case of a problem. If we had a second screen we might choose to place these graphics there.

FUNCTIONAL INFORMATION PROFILES: WORKING WITH EXISTING DESIGNS

In many cases, we don't have the luxury of doing a completely new design. We may want to enhance an existing design, or perhaps borrow a design concept from somewhere else. We can do this by comparing the

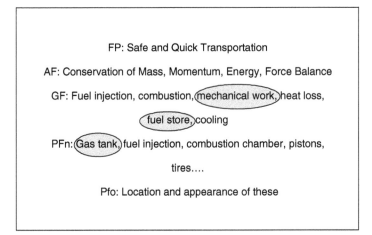

FP: Safe and Quick Transportation

AF: Conservation of Mass, Momentum, Energy, Force Balance

GF: Fuel injection, combustion, mechanical work, heat loss,

fuel store, cooling

PFn: Gas tank, fuel injection, combustion chamber, pistons,

tires....

Pfo: Location and appearance of these

Figure 4.7   Functional information profile.

existing design to our original work domain model through a functional information profile.

A functional information profile is a simple way of checking which areas of the work domain model have been covered and which have not. The original work domain model is used as a set of requirements. The existing display is compared to the work domain and mapped against the model. The exercise of creating the profile provides a method to the assessment. The profile at the end provides a summary of the results and indicates the degree of fit of the existing display to the model.

As an example, let's consider the typical variables on a car dashboard: speed, rpm, fuel level, and total distance traveled (odometer). If we look at these variables on the work domain model (Figure 4.7), we see that these variables provide us with some information at Generalized Function and possibly Physical Function. We also see that there are many areas that are not covered that may be amenable to design changes.

SUMMARY: FROM INFORMATION AND CONSTRAINTS TO DESIGN

If we review the example, we see that what may have seemed like an overly excessive analysis of an everyday system, a car, has led to some innovative graphical forms. Clearly some of the information is diagnostic and might be better suited to a technician, but the highest level graphics are feasible. They use information that is currently available on most cars (or could be added with a Global Positioning System). The main difference

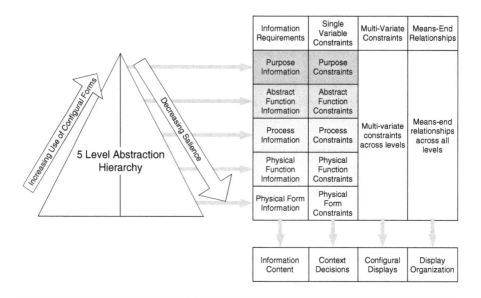

Figure 4.8   **Working from a work domain model to a design.**

is that they relate variables together, using graphical forms. In that sense, we have created a small ecological display.

In this chapter we have focused on linking our work domain model, from Chapter 2, with the graphical ideas from Chapter 3. We have proposed a four-step process for extracting information from your work domain model and using this information in your design. In our first pass we extracted all the information requirements from our work domain model. We used these to develop a very basic display. In our second iteration, we used our work domain model to determine constraints on single variables. We used these constraints to enhance the design and generate graphics that gave more information. In our third pass we extracted multivariate relationships from our work domain model. We used some of the more advanced graphical ideas in Chapter 3 to revise our design to show these relationships. Finally, we used the levels of the work domain model and the means-end links to organize our display and to determine salience levels between the graphics.

Design is an iterative process, and we have tried to suggest possible levels of iteration. In Figure 4.8, we summarize this design process. We recommend that your design take place in four main iterations, with these iterations looking at information requirements, single variable constraints, multivariate constraints, and means-end relationships. Each of these iterations should generate an improvement in your design.

5

TRANSPORTATION SYSTEMS

Transportation systems are primarily mass transport systems, moving themselves and some load or cargo to a desired place and time. These systems are more complex, though, and serve richer purposes than just moving things around. They have usually been designed for a specific context, be that passenger transport or military transport. The system is usually designed to transport people or goods safely; particularly with military vehicles, the vehicle may also have other capabilities such as being able to defend itself.

Transportation systems distinguish themselves from other kinds of systems in certain key ways. First of all, their primary purpose is to move themselves and a load from one place to another. This means that transportation systems are by their very nature not fixed. Second, they interact with the environment. This interaction is the second key way that transportation systems differ from other kinds of systems.

Transportation systems are similar to other systems because they have certain resources that they use. They are governed by the law of mass and energy; from that perspective, their models fit well with the approach of WDA. They are completely engineered systems, so the equations and relationships that govern how they work can be obtained.

In this case study we will discuss some of the key analytic and design challenges with transportation systems. We will show you some work domain models of systems, and compare and contrast different modeling options. Finally, we will show how these models have been used to generate design concepts.

CHALLENGES WITH TRANSPORTATION SYSTEMS

Analytically, the challenge with transportation systems is how to model something that moves freely in the world and can have multiple purposes.

There are several different options for modeling that will be discussed in the following sections. We will examine an analysis for a naval destroyer and a naval frigate that show two different ways of working through these challenges.

We will also look at how to handle the scale of your analysis under time constraints. In many situations there is only time for a partial WDA. This requires that the analysis be stopped at some point, while still incomplete. We will discuss how to choose this end-point in the case of the naval frigate.

When retrofitting a design on an existing system, there are often not enough sensors to support the design. In this case, we recommend that you conduct an instrumentation availability analysis (discussed in Chapter 2). We will show the WDA and instrumentation analysis for an aircraft.

Just as we don't always have the sensors we need, we don't always have the opportunity to build an entirely new EID. We will show partial EID concepts developed from these analyses.

In the following sections we will show you the analyses and discuss the major modeling decisions. Then we will compare the models in terms of how they handle the various types of challenges outlined above.

ANALYSIS FOR COMMAND AND CONTROL OF A FRIGATE

The analysis that follows was performed for the Department of National Defence, Canada. It was performed to explore WDA in the domain of naval command and control. The naval ship used in the analysis was the Halifax Class frigate, a multipurpose information gathering vessel.

Naval command and control is a domain that has been evolving with the emergence of new technology (Moorefield 1995). From a relatively centralized control structure, new naval command and control systems are becoming increasingly decentralized and distributed (Cebrowski and Garstka 1998). Access to new information and advanced decision support systems are increasing the range of decisions that an individual ship commander can make and making local distributed control more necessary. The nature of naval operations themselves has changed. Naval crews must participate in a variety of mission types, often facing less-organized and less-predictable operations other than war, and working within multinational task groups with different policies, skill sets, and equipment capabilities. Now, more than ever, decision support systems are needed that manage the increasing amount of information in a way that will create flexible, knowledge-based problem solvers from naval Command Teams.

The Canadian Navy is no exception to this situation. Canada's small-scale but technologically advanced fleet is frequently a member of multinational efforts. Its Halifax Class frigate was designed primarily for antisubmarine warfare, but in reality must take a role in a variety of missions, from surveillance to escort to information gathering, threat intervention, and the delivery of humanitarian aid. The Halifax Class frigate is equipped with a wide variety of above-water sensors, extensive sonar equipment, and, while not intended to be an offensive ship, enough weapon and decoy capacity to meet self-defense needs and to participate in a variety of missions. Its Operations Room Team works in a highly dynamic environment where multiple operators and controllers manage a large variety of incoming information. In the current Ops Room, this information filters and integrates upward through several levels of the team to the members of the Command Team, who manage the execution of the mission as a whole.

Many of the decisions made in naval situations are potentially critical and life threatening, to the ship itself and to society. There are underlying physical constraints of the situation that the team must operate within to be effective. For example, weapons have limited ranges and relations between range and the degree of accuracy that they can achieve. Ships have certain speeds that they can achieve and a fixed maximum maneuverability. Human operators are faced with the inevitability that they will eventually encounter unanticipated situations that will require flexible knowledge-based reasoning to manage effectively. There are several aspects of naval command and control that differ from other systems. A naval system has boundaries that are very difficult, if not impossible, to define. A naval frigate can be expected to undertake a variety of missions, varying widely in their types.

System Boundary

In this case, there were several possible boundaries that could have been chosen; for example, ship only, or contact only, or ship and environment. Making a selection on the boundary affected the scope of the eventual analysis and can affect the kinds of constraints that are revealed. For example, if the goal of the project is to support improved control over a certain subsystem of the ship, it makes sense to set the boundary at that subsystem and concentrate the analysis on only those details. However, if capturing interactions with external elements is of interest, a wider boundary may be necessary.

On this project we set the system boundary to include the ship but also to include the natural environment and contacts. We were interested in the interaction of ship capabilities with weather and sea features. Capturing information about contacts was an important role for the ship, so it made sense to include the contacts within the work domain model. We chose to leave out other friendly ships, air support, and land support. So, realistically, our modeling effort won't capture those constraints. We began the modeling process by developing a high-level model showing

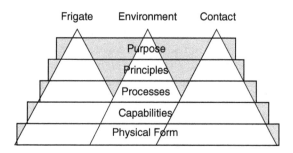

Figure 5.1   High-level work domain model bounding the system. (Reprinted with permission of Defense Research and Development Canada, courtesy of Bruce Chalmers.)

three parts of the work domain: the ship, its environment, and the contacts. Figure 5.1 shows this.

Work Domain Analysis

Following the general model developed above, work domain models were developed to expand each of the three areas. Figure 5.2, Figure 5.3, and Figure 5.4 show the models for the frigate, the environment, and the contact. These figures show more detail. You will see that the levels have been outlined and categories established. At Functional Purpose three main goals were described: movement, sea control, and survivability.

In determining Functional Purpose, it can be difficult to separate purposes from tasks. One way of distinguishing between purposes and tasks is to consider the length of time. While tasks are done for finite lengths of time, purposes coexist at all times. Purposes underlie tasks such that tasks work to achieve the purposes of the system. In the case of a naval ship, it performs many different tasks, but fairly few high-level purposes. We tested our purposes by generating a list of tasks to ensure that the purposes we defined covered the various tasks or missions of the ship. All tasks related back to the three purposes of movement, sea control, and survivability. Table 5.1 shows how we related ship tasks to purposes. Note that regardless of how different the various tasks or missions are, the purposes remain fairly consistent.

Abstract Function describes mass and energy, but also the expenditure of resources. A resource balance is a useful balance to include if resources are not changing form (e.g., food stays as food). The Generalized Function level models the basic processes, and the Physical Function level shows the major equipment types. But you will also notice that much more detail could be added to these models; for example, we could outline the various weapon and decoy systems and their individual capabilities.

Table 5.1   Mapping Missions (or Tasks) to Purposes

Task	Move from A to B on Time	Maximize Sea Control	Survive
Escort: Safe and timely arrival of vessels	x	x	x
Patrol: Detect and "deal with" all entities in patrol area	x	x	x
Screen: Safe passage of High Value Units	x	x	x
Surveillance: Gather all pertinent information within area of surveillance	x	x	x
Reconnaissance: Search area of reconnaissance and gather all pertinent information	x	x	x

In the frigate model in Figure 5.2, there are two types of elements shown. Physical work domain elements are shown in the white boxes. The gray boxes show social elements. Social elements constrain action but in a different way than physical elements. Physical constraints cannot be overcome; i.e., an engine only generates so much power, and a car can only go so fast. Physical constraints describe what you *can* do. In contrast, social constraints describe what you *should* do. They can be overridden, though; in many cases, like military environments, personnel are trained to respect these constraints as if they were physical laws. One way of handling social constraints in your model is to do two models. First do a work domain with only physical constraints. Then do a work domain with physical plus social constraints. Figure 5.2 shows this. It is important to differentiate the social constraints (as we did here by making them a different color). They should be marked as a reminder that, unlike physical constraints, it is still possible to break these constraints.

In the environment model in Figure 5.3, you will notice that the model only has four levels, beginning at the Abstract Function level. We opted to model the environment as purposeless and to concentrate the model on the lower levels of the analysis. At Abstract Function you see three laws modeled: conservation of mass, conservation of energy, and the creation of entropy (or the second law of thermodynamics). Natural processes create the Generalized Function level. At the Physical Function and Physical Form level there are many variables relevant to navigating a ship and performing in a naval environment.

In the contact model in Figure 5.4, you see a model very similar to the frigate model. In this case, however, the model is more general. The purposes are less specific, indicating only two purposes: some mission

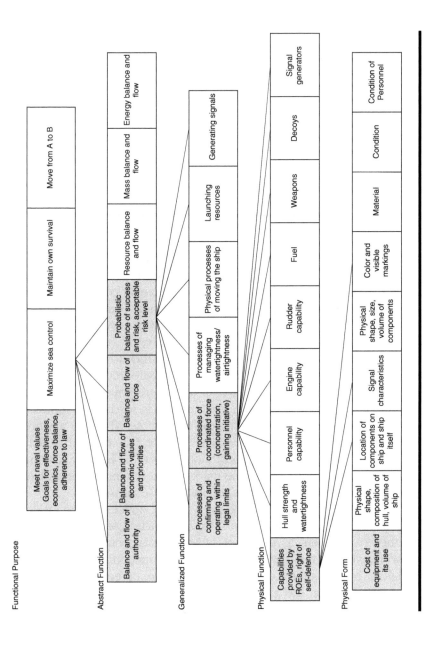

Figure 5.2   Work domain model for the frigate, showing physical (in white) and social constraints (shaded). (Reprinted with permission of Defense Research and Development Canada, courtesy of Bruce Chalmers.)

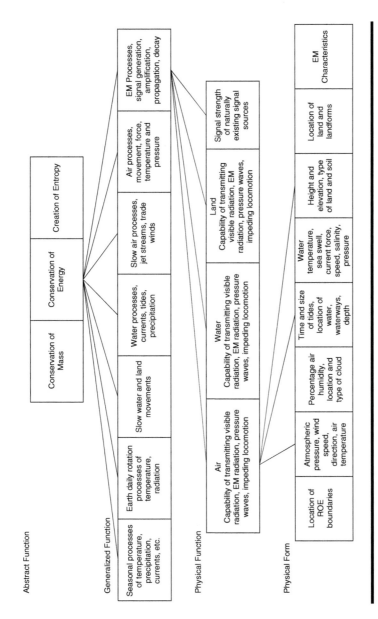

Figure 5.3 Work domain model of the environment. (Reprinted with permission of Defense Research and Development Canada, courtesy of Bruce Chalmers.)

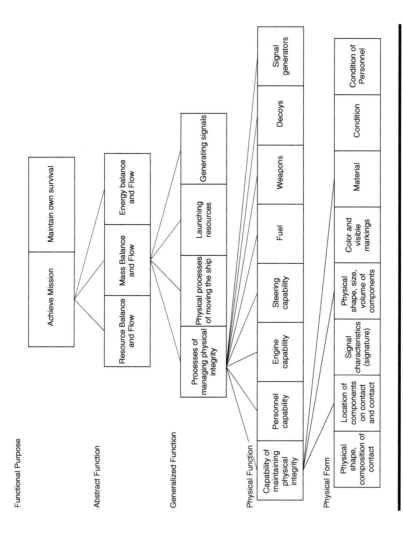

Figure 5.4 Work domain model of the contact, showing some means-end links. (Reprinted with permission of Defense Research and Development Canada, courtesy of Bruce Chalmers.)

and some survival. Capabilities and processes are also only general. The reason this model is so general is to allow it to be tailored later for specific contact types.

Testing a Work Domain Model

As mentioned in Chapter 2, it can make sense to test out your work domain model before continuing towards a design. Testing your model allows you to check for completeness, correct any misunderstandings, and keep in touch with your domain experts. In this project, the work domain models were tested by scenario mapping and a questionnaire evaluation. These methods are good for extracting rich information from a limited set of subjects, which is often the case when approaching domain experts. In this case, we will focus on demonstrating the scenario mapping technique.

In this case, we tested our work domain model by using a pre-existing naval scenario that had been adapted from a training exercise (Webb et al. 1998). Our objective was to confirm that, in stepping through the scenario, our work domain model captured all the domain constraints that were used in that scenario. If any constraints were missing from the models, we could improve our models by adding in those constraints.

The scenario was complex in that it involved nine different events with different levels of interaction, in different environmental conditions, day and night. The events ranged from sightings of unknown contacts, to warnings, to a single engagement (missiles), to a complex engagement of missiles and torpedoes simultaneously. Our scenario had air, surface, and subsurface contacts and occurred in a littoral (coastal) environment. Our domain experts were three Operations Room Officers (OROs) who are part of the Command Team of the frigate. They are largely responsible for handling the frigate in combat situations and second in command to the Commanding Officer, who has ultimate charge of the frigate. Our domain experts were familiar with the scenario from training, but not familiar with work domain models. They were given copies of the work domain models and explained the objective of the session and the purpose behind the models. We outlined the scenario step by step in a package and gave our domain experts code numbers for our work domain model elements to simplify transcription. Domain experts worked individually on each event, identifying which model elements were in use at each step and any other information that they use. At the end of each event, we held a discussion between all domain experts to compare opinions and extract more information with respect to the scenario.

As the event progressed, relevant areas of the work domain were highlighted, showing active regions spanning from purposes to Physical Form. Our work domain models included the models for the environments and for contacts. As a result, active areas in the contact of environments regions were also mapped out. We compiled the results using Powerpoint as our medium because of its ability to put the slides together to give a dynamic demonstration of work domain actions over time. Figure 5.5 to Figure 5.7 show static snapshots extracted from these files. The relevant elements have been expanded for legibility, while the remainder of the model is in the background. The models are available in Burns et al. (under review).

Figure 5.5 and Figure 5.6 show activities on the frigate side of the work domain. In Figure 5.5, an acoustic decoy generator is being prepared. The condition and readiness of the decoy is being checked, its potential interaction with other signal generators is checked, and power to the generator needed by the decoy is being brought on line. The purpose of launching a decoy is to meet the purpose of ensuring the survival of the frigate. Figure 5.6 shows the elements involved in launching a weapon. In effect, the weapon location changes, affected by weapon capability and the launch process. Launching the weapon decreases the resource store of the frigate, which is being used to maintain control over a certain sector of the sea.

Figure 5.7 shows a more complex interaction. In this situation, a contact has crossed a "Rule of Engagement" boundary in the environment, changing the types of actions that the frigate may take. The frigate team assesses the capabilities permitted in the new situation, their probability of success, and their authority in handling the situation relative to their naval objectives. Simultaneously, the capabilities of the contact, its movements, and its intended mission are all under consideration.

Evaluated together, the demonstrations created trajectories of reasoning across the work domain space over time. The primary purpose of these demonstrations was to show the completeness of the work domain constraints that were gathered. In no case did the domain experts mention constraints that were not in some way handled in the existing work domain models. It was noted that certain areas of the work domain were visited by the domain experts more often than other areas, and some areas not at all. This did not mean that these areas should not be in the work domain models, but rather that the scenario used here did not require the domain experts to enter those areas of the work domain. As an example, there were very few actions in the environment domain. A review of the scenario provided confirmed that very little environment information had been given in the training scenario.

In interpreting these maps we would make the following recommendations. First, as stated previously, a WDA component that is not used in this scenario should still be included in the analysis since it may be useful in other situations. That is, a validation exercise of this type indicates model elements that are missing, but does not necessarily capture every

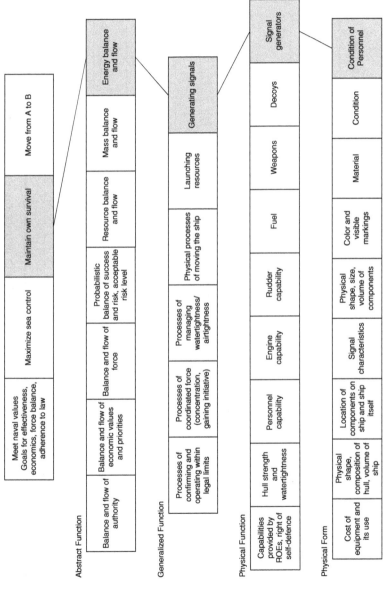

Figure 5.5 **Map of preparing an acoustic decoy signal. (Reprinted with permission of Defense Research and Development Canada, courtesy of Bruce Chalmers.)**

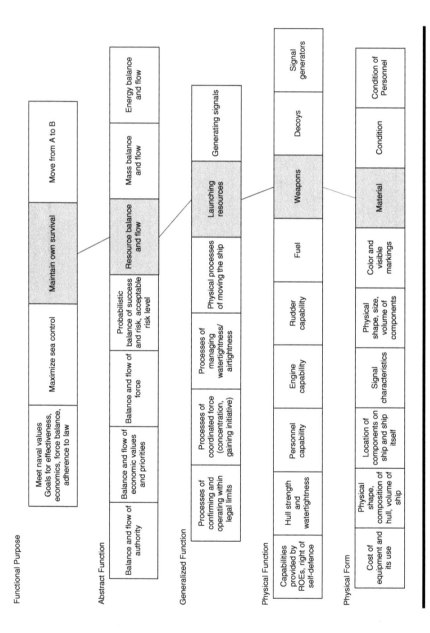

Figure 5.6 Map of launching a weapon. (Reprinted with permission of Defense Research and Development Canada, courtesy of Bruce Chalmers.)

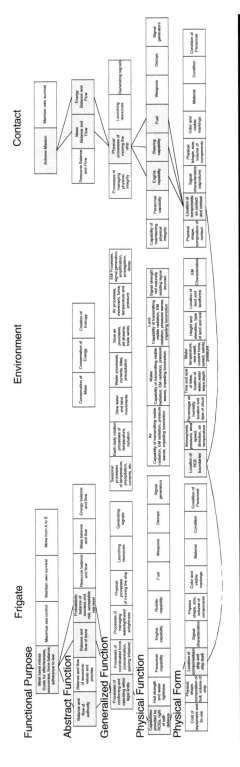

Figure 5.7   Evaluate contact's location relative to Rule of Engagement limits. (Reprinted with permission of Defense Research and Development Canada, courtesy of Bruce Chalmers.)

required element. Second, each time slice of the work domain shows concurrent relations. That is, an action at the lowest levels simultaneously activates the related elements at the higher levels; the directions of the arrows are merely to show relations.

The exercise of mapping a scenario against our work domain model served several purposes. First, it helped us to confirm and to validate that our model was reasonably complete. Second, the mapping of a scenario provided a good demonstration of the utility of our model. Researchers and domain experts familiar with task-based representations were able to appreciate the difference between task-based and work domain-based analyses, while demonstrating some of the connections between the two analyses. The approach added a third dimension to our WDA, in effect showing actions across the work domain over time. Furthermore, this mapping could be a useful linking step between WDA and later stages of Cognitive Work Analysis (in particular, Control Task Analysis).

ANALYSIS FOR COMMAND AND CONTROL OF A DESTROYER

The intent of this project was to conduct WDA to support the design of Watchstander tasks, functions, and support systems (e.g., controls and information displays) in the bridge and combat command center of a next-generation U.S. Navy surface combatant. Unlike the frigate project above, this was a futuristic ship that had not been designed yet.

Watchstanders are personnel who monitor and control the systems relevant to the ship's warfighting, defensive, and navigation functions including sensor and weapons systems. Two important constraints influenced the conduct of this project. Because the WDA was conducted at an early point in the ship's design, many details about both the general and specific types of physical systems that would be included in the ship's design were unknown. Additionally, the amount of time that could be allocated to collecting data regarding and creating the work domain models was limited. Further details on the methods and analyses can be found in Bisantz et al. (2003).

System Boundary

The system boundary in this project was defined as the ship and the battlespace (Figure 5.8). The battlespace can be considered an amalgamation of other contacts and the natural environment. The work domain model was done in a single model incorporating the entire battlespace.

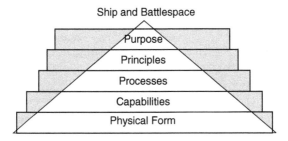

Figure 5.8   Bounding for destroyer analysis.

Work Domain Analysis

This analysis (Figure 5.9), though performed by different modelers, has many similarities to the frigate analysis presented earlier. At Functional Purpose, both mission-related and survivability purposes have been modeled. Rather than list specific missions, they have used a more general "Achieve missions" and then differentiated this between combat and noncombat mission types. This will allow the modelers to differentiate offensive purposes from defensive purposes (seen primarily in the Halifax analysis). Some work could still be done to define these purposes more clearly, but within the constraints of the project, this was a useful description.

Like the frigate model, the model includes both physical constraints and social constraints. Social constraints are modeled in more detail at the Abstract Function level in this model, and to a certain extent, physical constraints are not as thoroughly modeled. This reflects the stage in the design of the ship at the time of the modeling effort. This ship had not been physically designed yet, in contrast with the frigate that already existed. As the physical design became confirmed, this model would be reworked to outline physical constraints in more detail. At the Physical Function level, major systems are outlined with placeholders. The Physical Form level is not completed, since the ship does not exist yet. It is important to remember that work domain models can take several iterations to generate, and when working under tight project constraints, the models may also be constrained to fit the specific interests of the modeler and the project. As more information becomes available or time permits, the models can be improved. You will note that these modelers left a placeholder at Abstract Function for the addition of more physical constraints at a later time. They acknowledged that these constraints were there and left room for working them into the model later.

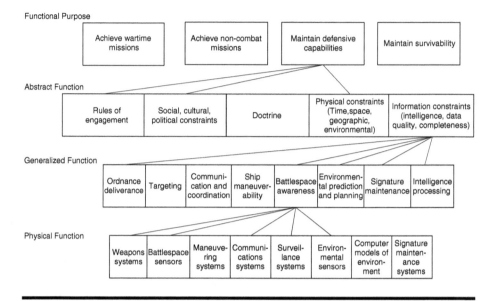

Figure 5.9   Work domain model for a destroyer, showing some means-end links. (Reprinted with permission from Bisantz, A. M., Burns, C. M., and Roth, E. M. (2002) "Validating methods in cognitive engineering: A comparison of two work domain models," *Proceedings of the 46th Annual Meeting of the Human Factors and Ergonomics Society*, 521–525. ©2002 by the Human Factors and Ergonomics Society. All rights reserved.)

A DISPLAY FOR AN AIRCRAFT

In this project the objective has been to use EID to enhance existing displays for an aircraft called the Harvard. The Harvard aircraft was a World War II plane used for training pilots. In this case, the particular aircraft was chosen because of access to the plane for eventually testing the displays.

The perspective of the project has been to concentrate on supporting commercial aviation, in contrast to military aviation. Therefore, offensive purposes or weapon systems did not need to be modeled. The other challenge of the project was to use EID in a partial implementation to enhance a currently successful display concept: the "Highway in the Sky" (HITS) display.

The Harvard is a basic early-stage aircraft without the complex automation of modern planes. It is a single-engine, propeller-driven plane with two fuel tanks. The basic components and flight principles behind the Harvard are similar to all current planes. Flight is achieved by using forward thrust over the various flight surfaces that convert thrust to lift. Controlling thrust and the shape of the flight surfaces through the elevators,

Figure 5.10   Classic interface of the Harvard aircraft.

ailerons, and tail rudder controls flight. Figure 5.10 shows the current classic Harvard interface.

System Boundary

The system boundary was mostly restricted to the plane with a small inclusion of environmental conditions, mostly windspeed. Other planes and air traffic were not included so that the analysis could focus on the description of the plane itself.

Work Domain Analysis

The plane was analyzed using the basic five-level work domain model. Levels of decomposition included the entire plane, the powerplant, powerplant subsystems, and subsystem components. The decomposition continued for hydraulics, the fuselage, and the electrical systems. Figure 5.11 shows the decomposition at the level of entire plane.

At Functional Purpose we have navigation and locomotion from point A to point B. At Abstract Function, the primary balance is the force balance that permits flight dynamics. The balances of energy and mass, while present, have not been shown in this high-level model. The lower levels are directed towards supporting flight, down to the curve of the wings modeled at Physical Form.

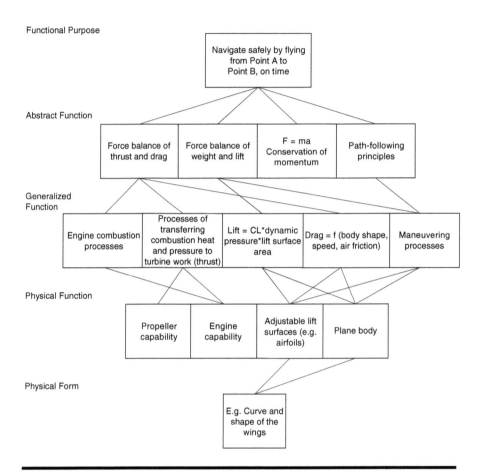

Figure 5.11   Work domain model at the level of plane.

At the next level of detail, shown in Figure 5.12, the four major systems are described. At this point, mass and energy balances are described as appropriate to the particular system. The major components are described for the hydraulic, powerplant, and fuselage systems at Physical Function. You will notice that the description has ended for the electrical system and the fuselage. They could be described further, but this was not of interest in this project.

The analysis continued, moving towards greater description of the elements.

Instrumentation Availability

This project plans to actually implement the displays on two platforms. The first test will be a simulation platform, which will use Microsoft Flight

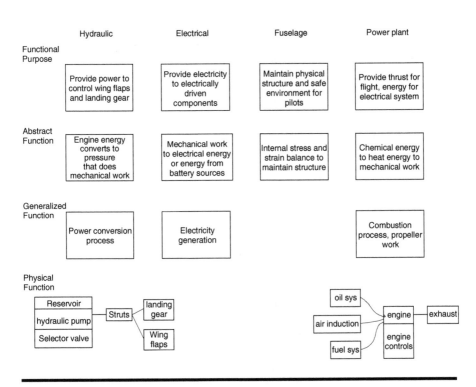

Figure 5.12   Work domain model at the level of systems.

Simulator with a Harvard aircraft dynamics and custom display package. The second test will be actual flight tests on a Harvard. Sensor addition or redesign options in both cases are quite limited.

For this reason we performed an instrumentation availability analysis for two situations. Table 5.2 shows variable identification with coding to show where the variables could be obtained. Note the missing variables at Abstract Function and Physical Form. These and many other variables would need to be calculated or derived in some other way.

Display Design

The objective in this project was to enhance an existing aircraft display, the Highway in the Sky display. The HITS display shows the corridor in the sky that the plane should be flying in and has been successful in improving navigation in experimental tasks. The objective was to strategically improve the display, not to provide a complete EID display.

The HITS display shows the location of the plane in three dimensions: altitude and ground-based x and y (or longitude and latitude) relative to a referent, which is the desired flight path. In controlling the plane, however, the WDA shows that speed and thrust control will interact with

Table 5.2   Instrumentation Availability for the Harvard Aircraft

Level	Available Variables	Aircraft	Flight Simulator
Functional Purpose	Heading	x	x
	Engine power		x
	Propeller power		x
Abstract Function	No variables		
Generalized Function	Velocities	x	x
	Accelerations	x	x
	Engine oil pressure		x
	Engine temperature		x
	Manifold pressure	x	x
	Fuel flow rate		x
	Oil pressure		x
	Piston pressure		
	Cylinder pressure		
Physical Function	Pitch	x	x
	Roll	x	x
	Yaw		
	Angle of attack		x
	Altitude		x
	Propeller rpm		x
	Fuel temperature		x
	Oil temperature		x
	Engine torque		x
	Piston state		
	Cylinder state		
Physical Form	No variables		

altitude control (Abstract Function and Generalized Function levels). The analysis also showed that the purpose of the plane was to arrive at point B at a certain time. This suggested that the HITS could be improved by adding airspeed and climb rate constraints and information on the time and distance to destination. While the analysis revealed many other variables, they did not connect directly with HITS enhancements and so were not used in this project.

Figure 5.13 shows a very basic HITS display showing the pathway and location of the plane in x, y, and z.

Figure 5.14 shows EID enhancement of HITS; in this case, a heads-up HITS is visible with the red rectangles to the right of the photograph. Three EID-based enhancements are visible: a triangular distance-time navigation tool, an airspeed trend, and a glideslope indicator.

Figure 5.13   Basic HITS display.

EYE ON VISUALIZATION

This display (Figure 5.14) uses:

1. A configural triangle to show speed, time, distance
2. A vertical line as an emergent feature for on-time status
3. Two different trend plots
4. Background shadings to give context on the trend plots

THE ANALYSIS IN THE DESIGN

Analysis: Figure 5.11 and Figure 5.12

This display only uses a small part of the analysis. The key features are:

FP: Speed time and distance goals   The triangle graphic
GF: Airspeed and airspeed rate of change   Airspeed trend chart
PFn: Altitude constraints on landing   Glideslope trend with
                                               constraint region

Figure 5.14   The EID-enhanced HITS. The enhancements are on the main cockpit.

Evaluation

This display was tested with six pilots and six gameplayers (Moradi-Nadimian 2003). The task was to fly a predetermined flight path and land at a target airport at a designated time. In different scenarios, pilots experienced different visibility and wind levels, and some scenarios had instrument failures. In all cases, the enhanced display generated more accurate landing times, with the most noticeable improvement in the most challenging weather conditions.

An Alternative Design: Integration with HITS

Amelink (2002) independently worked on a similar project, aiming to use EID to enhance HITS support for a flight simulation of a Cesna Citation 500. Amelink performed a WDA and then focused on the energy conversion from potential to kinetic energy. There are many similarities between the work domain models of Moradi-Nadimian and Amelink; in fact, on

Table 5.3   General Work Domain Model behind Amelink's Design, Adapted Slightly from Amelink et al. (2003)

Functional purpose	Fly trajectory (follow speed profile and altitude profile)
Abstract function	Law of conservation of energy
Generalized function	Energy management, control of kinetic and potential energy to make total energy
Physical function	Effect of components on state variables and aircraft control
Physical form	Aircraft-specific components and their configuration

major points they are identical. Amelink further supported his modeling effort by doing extensive control task modeling of energy management.

From this analysis he developed an enhanced HITS display showing energy requirements, shown in Figure 5.15. One of the key differences between this display and the previous one is that this display fully integrates the speed control information with the HITS in a very clever design.

The information added by this display is quite similar to the previous display — both displays have chosen to add airspeed, climb rate, glideslope,

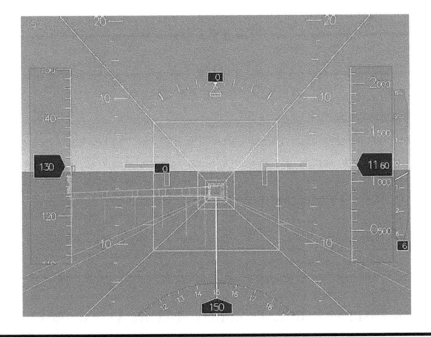

Figure 5.15   Amelink's energy bar graphic showing a deviation from the demanded speed. (From Amelink, M.H.J., Unpublished thesis, 2002.)

and groundspeed information. The previous display uses four separate graphics to add this information. Amelink's display integrates the information together. Amelink's display should provide some exciting control improvements, making control actions quite perceptible on the display. So far, though, the display has not been tested very thoroughly. The display was tested with two pilot participants who flew four flight tasks with atmospheric turbulence. NASA TLX workload measures were collected as well as subjective experience measures. This relatively brief study showed mixed results for the enhanced display, possibly reflecting the extensive prior training of the pilots and the difficulty of changing over to such an innovative display.

EYE ON VISUALIZATION

This display (Figure 5.15) uses:

1. Extensive information integration

2. Triangular geometric relations on the speed control features (yellow lines)

3. A horizontal emergent feature (the green target and green line)

4. Semi-transparent meters as gauges

THE ANALYSIS IN THE DESIGN

Analysis: Table 5.3

This display only uses a small part of the analysis. The key features are:

FP:	Speed and altitude goals	Demand speed and altitude indicators (on green and yellow profile)
GF:	Energy relationships	The green speed profile and yellow altitude profile work together
PFn:	Speed and altitude	To display how speed and altitude affect kinetic and potential energy

---

A DISPLAY FOR BALANCING FUEL IN AN AIRCRAFT

As an example of EID performed on a subsystem, Dinadis and Vicente (1999) conducted a WDA that was conducted for the engine and fuel

Figure 5.16 The traditional control panel for balancing fuel on the Hercules.

system of a Hercules transport aircraft. From the analysis, a novel display was developed. You will see, at this level, the greater detail that arises with this more contained analysis.

The Hercules aircraft is used for military transport. Its fuel balancing system that feeds the engines is fairly complex, as can be seen from the photo in Figure 5.16. The aircraft has two fuel tanks that feed four engines. The tanks can feed the engines in several different ways, and pilots must monitor the levels of the tanks, ensure enough flow to the engines, and balance the tanks to balance the weight of the aircraft.

System Boundary

The boundary in this project included only the engines, the fuel tanks, and the equipment within the general engine system (such as the fuel heater, oil store, and oil transport). Other parts of the aircraft were not part of this project.

Work Domain Analysis

A very high level WDA for this sytem is shown in Figure 5.17.

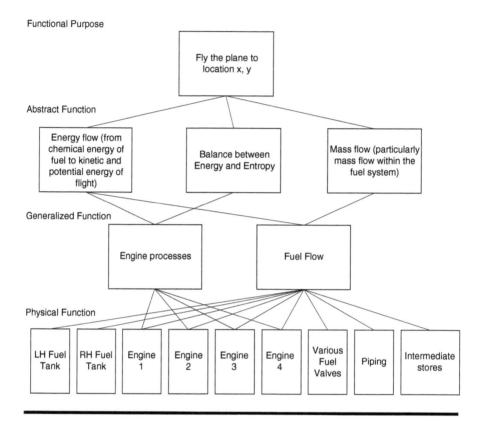

Figure 5.17   A high level view of the work domain model for the fuel balance display.

Display Design

Several key aspects were derived from the WDA that went into the EID. First, from the Functional Purpose level, the overview display on the left was designed. This display shows the distance that can be reached with the current amount of fuel: a direct translation of Functional Purpose information.

Across the top of the display there are four polar star displays, one for each engine. The polar star displays integrate multiple variables from the engines (such as rpm, operating temperature, and so on). These are Generalized Function variables. The visual integration into the shape of the polar star adds new information on the overall status of the engine, or Functional Purpose information at the engine level of decomposition. The four pink graphics below the polar stars show Abstract Function information and relate the energy and entropy of the engines at different stages of the engine cycle (e.g., compression, combustion, and expansion).

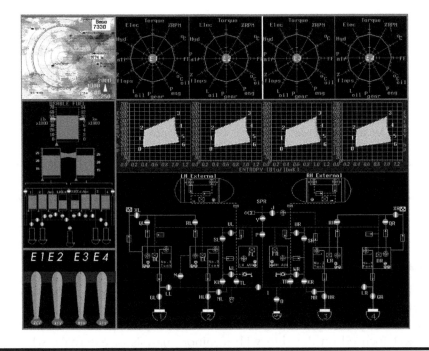

Figure 5.18 The fuel balance display. (From Dinadis, N. and Vincente, K.J. (1999) International Journal of Aviation Psychology, 9:241-269. With permission from Lawrence Earlbaum.)

On the left of the display, in blue, is the fuel balance. This display shows the overall fuel amount, the balance across the two tanks, and the contribution to the four engines. Valve settings, open or closed, are also shown. This particular display shows two levels of information. The balance is a concept derived from the Abstract Function level of the WDA. The actual tank levels and valve settings are from the Physical Function level.

EYE ON VISUALIZATION

This display (Figure 5.18) uses:

1. Four polar star displays for engine state
2. A map as an overview, showing how far the plane can fly
3. Connected bar graphs for fuel balancing
4. Highly salient process graphs of engine behavior

Analysis: Figure 5.17

This display is a full WDA implementation. The key features are:

FP: Flight distance goals   Map showing flight range in top left corner
      engine state goals   Four polar stars of engine state
AF: Mass balance relationships   Left side balance graphic
GF: Engine processes   Power over engine stages in central region
PFn: Components   Mimic diagram at bottom right of display, con-
                                             trols bottom left

HANDLING THE CHALLENGES

Challenge 1: Determining the System Boundary

In these four examples we have seen different system boundaries. On the Hercules, the boundary was drawn around the fuel system, in contrast to the Harvard, where the analysis focused on the whole plane. The destroyer project focused on the ship with a minor analysis of externals, whereas the Halifax project strongly included external entities within the work domain. In Table 5.4, we show the project objectives and how the system boundary matches with those objectives.

Table 5.4   Matching Objectives with System Boundaries

Objective	Boundary
Improve fuel balancing	Fuel and engine system
Improve navigation	Entire plane
Determine ship requirements	Ship with some externals
Manage ship resources and situation and threat assessment	Ship, environment, and contact

As we mentioned earlier, system boundaries are artificial. Setting them determines the scope of the analysis, but it can also affect the quality of the analysis. A system boundary that is too broad will have the analysts wasting time on sections of the work domain that are not relevant. With time and money constraints, this can prevent them from reaching a deep enough level of detail in the parts of the work domain that are really of interest. Similarly, a boundary that is too tight will restrict you from collecting all of the necessary information. You should be able to notice the difference in detail between the Hercules analysis, which describes fairly small engine components, and the Halifax analysis, which describes only "Engine systems." If the engine systems were of key interest in the Halifax project, this analysis would have to go much further. On the other hand, the current Hercules analysis would be too restrictive to understand certain problems, like how the aircraft navigates or interacts with other military units.

Challenge 2: Determining Purpose

When doing a WDA, determining the purpose of the system is key. This is one of the aspects that distinguishes WDA from task analysis. Tasks change hourly, daily, or monthly, but the purpose of the system should remain constant throughout.

In the Halifax example, Table 5.1 showed a useful way to transition from tasks to purposes. In many cases it is easier to first think of tasks, since tasks are what we work on every day, and task performance is clearly something we want to improve. Thinking about purposes requires thinking through and beyond the level of tasks and determining what remains constant throughout the tasks. If we can improve these constants, we should be able to improve task performance regardless of the task. Remember, tasks are always changing and evolving, so it makes sense to start your design by aiming to improve purposes.

If you are having difficulty determining purposes from tasks, the transition table shown on the Halifax is a good strategy to use. List all the tasks that you can think of in one column and then start to determine what is constant among them. List your proposed purposes across the other dimension and work through your tasks to see whether or not your purposes hold for most of your tasks. You will have a powerful analysis by determining these constants. By example, contrast the concepts of "Maximize sea control" from the Halifax project with "Achieve wartime missions" from the destroyer project. While "Achieve missions" is a way of grouping tasks together, this description does not actually reveal what constants lie across the various mission types. Alternatively, the concept of sea control begins to reveal that ship equipment, from weapon systems to information gathering equipment, exerts some sort of influence over other contacts over some range of distance and time. This constant holds across many different mission types.

Most systems are designed for only a couple of purposes and are more straightforward than a naval frigate. Once you have determined purpose, it is important to express it carefully. You want to express purpose as specifically as possible and try to look for at least two purposes. If we look at the Harvard analysis, the purpose has been described as "Navigate safely from point A to point B." While this purpose is specific, there are actually two purposes expressed here — one to navigate from point A to point B, and the other to navigate safely. The analysis could be improved by differentiating these. For an example of separating these purposes, consider the Halifax example. In this analysis, purposes are described as "Move from A to B" and "Survive." This can allow you to separate safety systems from operational systems and to examine any performance conflicts that may arise from operating safely.

Challenge 3: Modeling Physical Relationships

Physical relationships fit easily into WDA. Looking across these examples you can see a great deal of consistency in how the models were developed. If you compare level by level you will see relatively similar descriptions, varying mostly in the language or detail that was used. All of these systems model mass and energy relationships at the Abstract Function level. The Harvard project goes one step further and models the force balance as well, a balance that is relevant for describing the flight of the plane. At Generalized Function you find different types of processes; for example, "Combustion" on the Harvard aircraft, the engine processes on the Hercules, the "Physical processes of moving ship" on the Halifax. Physical Function contains a description of the components and Physical Form of those components.

Challenge 4: Modeling Social Relationships

Social relationships show a different kind of constraint. If you interview operators you may find that they consider these constraints as solid as physical constraints and genuinely use them to guide their behavior. It makes sense, therefore, to model these in the same way using the WDA. Table 5.5 shows the social constraints that were modeled in the Halifax and the destroyer examples. The aviation examples did not model any social constraints.

If you examine both naval examples, you will notice that social constraints are more apparent at the higher levels of the models. This is because social constraints reflect values, priorities, and possibly certain procedural ways of meeting those values. Social constraints are less relevant when describing the physical capabilities or attributes of equipment, although equipment costs and capabilities within rule sets may be possible expressions of social constraints at this level. Generally, however, social constraints add an additional level of constraint on the use of your resources, and this is why they are reflected at the higher levels, and not as much at the actual resource-based levels of Physical Function and Physical Form.

Table 5.5   Social Constraints from the Frigate and Destroyer Models

Level	Intentional Constraint
Functional Purpose	Meet naval values
Abstract Function	Economic values and priorities
Abstract Function	Acceptable risk level
Abstract Function	Social, cultural, and political constraints
Generalized Function	Processes of operating within legal limits
Generalized Function	Communication and coordination
Physical Function	Capabilities provided by rules of engagement and rights of self defense
Physical Form	Costs of equipment and their use

We suggest, though, that you use some method of distinguishing these constraints from physical constraints so that you can look at your model in two ways: with only the physical constraints and with the physical and social constraints. The Halifax example shows one way of doing this. In this project, the physical models were developed first and then the social constraints added to them and identified with different shading. With this kind of approach, the analyst can move between the two different models and differentiate between laws that physically cannot be broken, and those actions that really should not be taken but are still physically possible. In contrast, although the destroyer model includes social constraints as well, they have not been separated from the physical constraints.

Challenge 5: The Number of Levels and Models

In general, five-level models are the norm in WDA. The five-level model arises from the different types of information that are required: purposes, principles, capabilities, and so on. In general, we recommend that you work towards a five-level model in most cases. There are instances in which a different number of levels have been used, and you see this in both the Halifax and the Harvard examples.

In these examples, though, the levels still exist; the modeler, however, has not completed the level for various reasons. In the Halifax you see that Functional Purpose on the environment side has not been completed. This is to reflect that, unlike the ship and the contact, the environment is not a designed entity. It therefore acts without purpose, although it still adheres to physical laws and has capabilities. Depending on the perspective, there may be a way to describe this level, although further description was not relevant for this project.

The Harvard example shows the other situation where a level is not completed. In this project, certain levels were not completed largely because they were not of interest. The lower levels of the electrical system show this. If the project scope included the electrical system, then these levels should be completed. The levels are left blank in the model to show that they still exist and are just not described. When leaving levels empty in your analysis, you should think about why you are not describing that level and record your justification.

It is rare that a six-level model is required, and we have not been able to find any strong examples of a six-level model. Before continuing with a six-level model, you should review your model and make sure that one of your levels is not just a grouping or an aggregation of one of the other levels. It is easy, but not desirable, to integrate a level of Part-Whole Decomposition into your model.

As well as the number of levels, you may have to decide the number of models that you will produce. In general it makes sense to develop one model for each level of Part-Whole Decomposition, including one model for each subsystem or component. This generally comprises a complete analysis. There is no reason, though, that you cannot subdivide your models further, and, indeed, with very complex systems this is necessary in order to keep the models readable.

In the Halifax example, we showed the technique of dividing the work domain by another dimension besides the Part-Whole Dimension. This example divided the work domain by entities (ship, environment, and contact) and developed separate models for each entity. In contrast, the destroyer example, for a very similar system, kept everything within one model. If you compare the two examples you will see the advantages and disadvantages of these approaches. If we keep everything together in one model, we can easily see the connections between all the elements. Note that in the destroyer example, "Battlespace awareness" is clearly connected to "Physical constraints." The equivalent part of the model in the Halifax model is the contact model, since understanding the contact is the equivalent of battlespace awareness. The connection to the physical capabilities of the ship is not as clear, but the model has more detail in this section. In particular the purpose or intention of the contact is modeled. This shows the trade-offs between the two approaches: one shows closer connections, and the other shows more detail.

Opportunities to divide the work domain arise under certain situations. If you have multiple interacting entities it may make sense to do a model for each entity. If the different entities have conflicting purposes, this is also a good indicator that separate models may be required. The other situation when you may wish to divide your work domain model is when you have distinctly separate resource groups working towards a similar purpose. This occurs in accident response and mitigation domains. The domain of the accident and the domain of the resources to mitigate the accident are most likely distinct and use different processes and principles. You may find it useful, therefore, to model these parts of the work domain separately.

Challenge 6: The Depth of Analysis

The required depth of analysis is a challenge in every project, and throughout this book you will find several examples of different decisions. Depth of analysis is practically constrained by the time and money devoted to the project, and this is seen in the Halifax and the destroyer examples. In both cases, placeholders have been used to indicate where further analytical work could occur. The destroyer example also shows a different constraint on the depth of analysis. In this project, very little information was available on the physical characteristics of the ship. Realistically, these parts of the model could not be developed further at the time, so again, placeholders for the physical characteristics have been added.

In comparing the Harvard and Hercules projects, we see the depth of analysis constrained by the needs of the project. The Hercules project requires that the analysis proceed down to the smaller components within the engine system. In contrast, the Harvard example stops at the general definition of "engine." Whereas the Hercules project is directed at improving fuel control for the aircraft, the Harvard is focused on more general flight path control and, at this point, does not need that level of detail. A way of determining the level of analysis comes from looking at the level of control required. Your last level of your WDA must show the control that you need. You can observe how the depth of analysis and level of control influences the design of the displays as well. Notice how the Hercules display shows the valve settings and control levers for the fuel system. The Harvard is much more of a flight monitoring display and does not show specific equipment status.

Challenge 7: Analysis at Different Design Stages

The four examples in this chapter show work at different stages of design. The Harvard and Hercules projects examined existing designs that were stable; that is, not undergoing design changes. A lot of detail can be developed in these models. In contrast, the destroyer project shows an analysis for a ship that was just being designed. In many cases, information and specific details were not available. The analysis had to focus on defining purposes and high-level priorities.

The Halifax example shows a third alternative. In this project, the ship was in existence but was undergoing a retrofit of many of its systems. So while a lot of detail was available, it made sense for the analysis to focus on the environment and contact models. These models in themselves could be used to examine the weapon and sensor capabilities needed by an upgraded ship. In effect, the operators of these ships need to know all the information in the environment and contact models to act properly in their work domain. Any information in these models that was not available in the current control center showed an area for improvement. These models, therefore, provide a useful definition of information needs. This shows another reason to divide the domain into different entities. The added level of detail can be used to define new sensor and equipment needs for your system.

Challenge 8: Instrumentation Availability

In the previous challenge discussion we talked about defining new sensor requirements from the WDA. These requirements can come from the instrumentation availability analysis. The instrumentation availability analysis (for example, on the Harvard) is a simple map of which variables from the WDA are available in the system. This mapping can be used in several different ways. You may be only able to implement a design using currently available information. But in many cases you may be able to add new information by calculating information or designing new sensors. This makes the information availability analysis a very useful tool.

The Harvard project is using this analysis in yet a different way. Since the project will test displays on a simulator as well as in real flight, the analysis allows us to systematically compare the information that is available under the different conditions. So in this way, multiple information analysis can be conducted from the same WDA, if the displays are going to be implemented on similar systems with different levels of sensor support.

Challenge 9: Partial Display Implementations

In reality, many domains already have existing display sets that are performing well, are thoroughly tested, and are well accepted in the domain. Presenting a completely new design in these domains is not always well accepted. WDA can still be used to develop concepts that can enhance an existing design. The aviation examples of the Harvard and the Cessna both demonstrated this idea.

There are several strategies for choosing an effective partial implementation:

1. Check that all relevant Functional Purpose information is being displayed. Many systems only display primary purposes related to production or business goals. Safety, financial, and maintainability purposes are rarely shown.
2. Check that Abstract Function and Generalized Function information is being displayed. These are the two levels least likely to be found in existing displays.
3. Look for challenging control tasks. Amelink's display was driven by the task of controlling speed. See Vicente's book, *Cognitive Work Analysis* (1999), for more information on performing a control tasks analysis.

Partial implementations can either support the existing display with the additional information in another view area (e.g., Moradi's design) or be developed directly into the existing display (e.g., Amelink's design).

6

PROCESS CONTROL SYSTEMS

Process control systems are used to make things under conditions or reactions that require tight and ever-present control. Examples of process control systems are power generation from coal or nuclear sources, petrochemical production, and oil refining. Some manufacturing and food processing systems can also be considered process control systems. These systems involve moving different substances, harnessing and controlling energy sources, and in most cases controlling processes or reactions. The system is usually designed with some sort of end product in mind and may also have requirements to operate safely, to operate efficiently by minimizing the use of energy or materials, and to operate within environmental regulations. These requirements can change daily, and the process itself may experience disturbances that require operator intervention. These systems can involve automation but are usually of such a scale and complexity that total automation has not been feasible.

While great advances have been made with technology, process control industries still see a significant number of incidents and waste in process operations. It has been estimated that $10 billion per year is lost in the United States alone due to these incidents (Nimmo 1995). Many of the cases could have been improved by better control and more rapid and accurate operator response.

Process control systems distinguish themselves by the nature of the processes they control and the end products they must produce. Mass and energy relations are typically important in controlling these systems, as is an understanding of the physics and chemistry of the processes, their feeding products, and their end products. These systems are well handled by an EID approach; indeed, the EID approach originated from the study of these systems. Chapter 1 reviews the foundation work on EID, in particular the DURESS example. DURESS was a simulation of a feedwater control system that has been used for many studies behind EID. In this chapter we will focus on expanding the concepts developed from DURESS to larger systems.

CHALLENGES WITH PROCESS CONTROL SYSTEMS

The challenges we will discuss in this chapter include:

- What to do when there are issues with instrumentation or sensor availability
- Alternatives in part-whole representation
- When to use functional, causal, or other models
- How to organize information on an EID display
- How to add task analysis information to an EID display

Process control systems are defined by key process relationships. These relationships can be modeled in the WDA and translated into effective graphics. Due to the scale of these systems, there are often issues in terms of how extensive the analysis should be, how to handle information requirements that can't be sensed, and how to distribute information effectively across different screens. The following cases have been selected because they demonstrate ways of handling these issues effectively.

CHALLENGES WITH PROCESS CONTROL SYSTEMS

Dealing with instrumentation or sensor availability
Using a part-whole representation
Using functional, causal, or other models
Organizing information on the display
Adding task analysis information to an EID

DESIGN FOR THERMAL POWER GENERATION SYSTEM

This project involved the design of ecological displays for a simulation of a thermal power generating plant. The simulation contained main components and sensors and is shown in Figure 6.1.

System Boundary

The system boundary included the components in Figure 6.1.

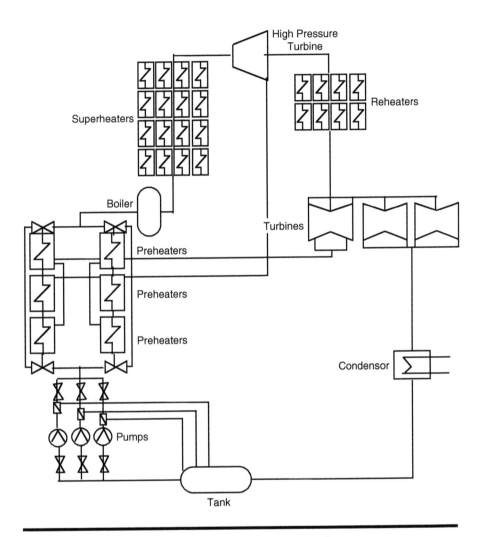

Figure 6.1 General design of the thermal generating plant.

Work Domain Analysis

Figure 6.2 shows a high-level WDA. This figure shows the regions that were modeled and general types of information in each region. Only four levels were modeled because this design was for a simulation. The model identified mass and energy relations at the Abstract Function level. The Generalized Function level revealed a cycle of water-to-steam changes known as a rankine cycle, which is typical of power generation systems that use steam turbines. The number of components and sensed variables was moderately high, 402 variables, reflecting the complexity of this

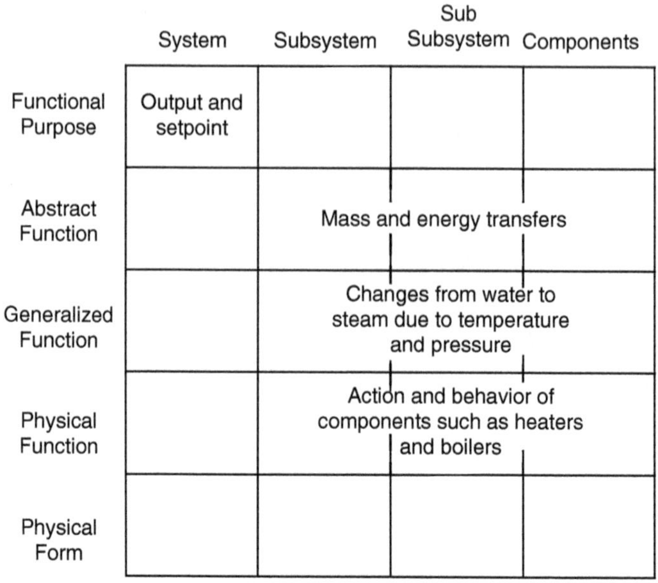

	System	Subsystem	Sub Subsystem	Components
Functional Purpose	Output and setpoint			
Abstract Function		Mass and energy transfers		
Generalized Function		Changes from water to steam due to temperature and pressure		
Physical Function		Action and behavior of components such as heaters and boilers		
Physical Form				

Figure 6.2 Work domain model of a coal-fired power plant, overall view. (Reprinted with permission from Burns, C. M. (2000) Putting it all together: Improving display integration in ecological displays, *Human Factors*, 42:226-241. ©2000 by the Human Factors and Ergonomics Society. All rights reserved.)

implementation. Figure 6.3 shows a slice of this model at the subsystem level. Note how the Functional Purpose level has not been modeled, but there is more detail on Abstract Function and Generalized Function levels. Figure 6.3 is also an example of a causal Abstraction Hierarchy model; it shows the general flow pattern.

Display Design

When displays were designed for this system, three different design options were considered. With the relative scale of the plant, there were several options for displaying the information. The information could be displayed on multiple screens, shown in multiple windows, or integrated into a single but complex display. Figure 6.4 shows the different design alternatives. In each case the acronyms stand for the different WDA levels: Functional Purpose (FP), Abstract Function (AF), Generalized Function (GF), Physical Function (PFn). The option on the left shows the situation where each level of the WDA is on a separate screen; the option in the middle is where each level is in a separate window on the same screen; and the third option is where all four levels are on the same screen, tightly integrated.

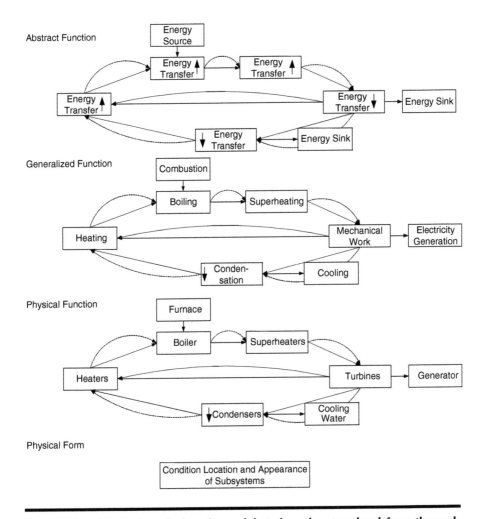

Figure 6.3 A causal work domain model at the subsystem level for a thermal generating system.

A display was designed for each WDA level, and at each level there were three levels of detail matching the part-whole levels shown in Figure 6.1. The design of the displays was driven by the need to develop the third, integrated option. This meant that the layout of the display elements was organized to meet the constraints of the integrated design. The various display items were then lifted off as different layers by WDA level to create the separate screen and the windowed options. This resulted in the displays in the next three figures. Figure 6.5 shows the integrated display; Figure 6.6 shows the windowed display; and Figure 6.7 shows the display with one level at a time, in the Physical Function view.

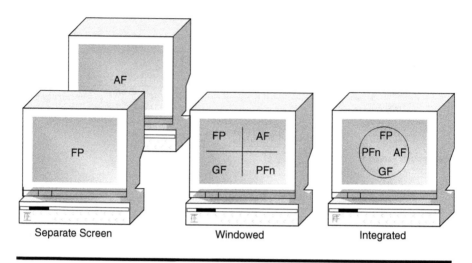

Figure 6.4   Design alternatives for navigating through a large ecological information space.

Graphic Form Design: The display in Figure 6.6 uses several iconic and configural displays. The iconic tank and heaters also double as analog displays and fill as the component fills. The Generalized Function level shows a configural graphic of temperature, pressure, and entropy. This kind of plot is known as a rankine cycle plot and reflects the characteristics of water. It was first proposed as a display for process control by Beltracchi (1989; 1995) and evaluated successfully by Vicente et al. (1996). The Abstract Function layer uses a bar graph of mass and energy levels with a configural pattern to the data. The pattern is emphasized by the gray contour behind the figure.

Evaluation

The three displays were evaluated using undergraduate engineering students at the University of Toronto (Burns 2000) and then re-evaluated at the University of Waterloo (Burns et al. 2002). These evaluations have found:

1. Slower fault detection times with the integrated display
2. Faster fault diagnosis times with the integrated display
3. More accurate fault diagnosis with the integrated display
4. Generally poorer results with the windowed display
5. Higher workload with the windowed display

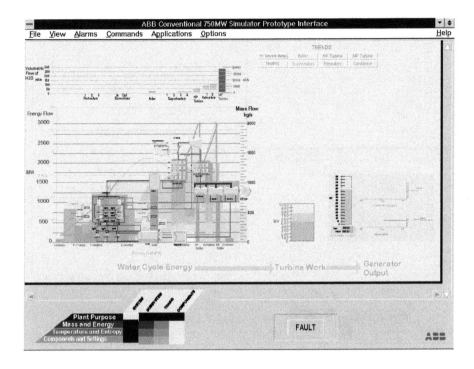

Figure 6.5 Integrated view. (Reprinted with permission from Burns, C. M. (2000) Putting it all together: Improving display integration in ecological displays, *Human Factors*, 42:226-241. ©2000 by the Human Factors and Ergonomics Society. All rights reserved.)

DESIGN FOR A NUCLEAR POWER SIMULATION

In this case study, Yamaguchi and Tanabe (2002) used EID to generate displays for a nuclear power simulator. The design also adapts ideas from the DURESS design and applies them to a larger system.

The work environment in this case is an engineering simulator for the nuclear ship Mutsu. The ship has a two-loop pressurized water reactor with a core power of 36 MW. This reactor system was designed to supply power for the ship. In the system, coolant water is circulated by main coolant pumps and is heated by the reactor core. The heat from this water is transferred to a second loop that moves steam through the turbines in order to produce energy. The turbines drive the ship and provide electrical power for the ship. The two-loop structure is used to provide containment of reaction products in the reactor loop and to isolate these products from the other components in the system. The reactor simulator can generate normal operating conditions and different kinds of malfunctions.

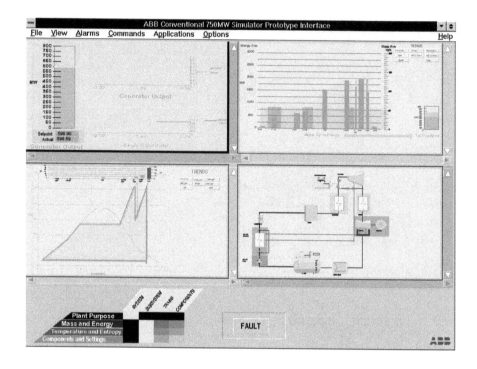

Figure 6.6 Windowed view. (Reprinted with permission from Burns, C. M. (2000) Putting it all together: Improving display integration in ecological displays, *Human Factors*, 42:226-241. ©2000 by the Human Factors and Ergonomics Society. All rights reserved.)

System Boundary

The system boundary was restricted to the reactor and the two water loops. The main reactor loop was included, as was the loop responsible for transporting steam through the turbines. Auxiliary and other systems are not included in this example.

─────────────────▼▲▼─────────────────

THE ANALYSIS IN THE DESIGN

Analysis: Figure 6.3

This display is a full WDA implementation. The key features are:

FP: Power demands   Goal meter in top left corner Figure 6.6
AF: Mass and energy balances   Bar graph in top right corner Figure 6.6
GF: Rankine cycle   Graphic in lower left corner Figure 6.6
PFn: Components   Mimic diagram in lower right corner Figure 6.6

─────────────────▲▼▲─────────────────

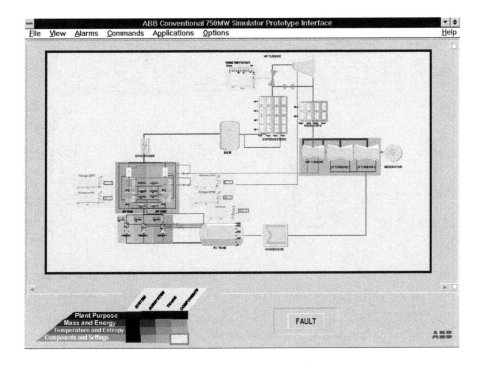

Figure 6.7 Physical Function display. (Reprinted with permission from Burns, C. M. (2000) Putting it all together: Improving display integration in ecological displays, *Human Factors*, 42:226-241. ©2000 by the Human Factors and Ergonomics Society. All rights reserved.)

━━━━━━━━━━━━━━━━━━━━ ▼▲▼ ━━━━━━━━━━━━━━━━━━━━

EYE ON VISUALIZATION

This display (Figure 6.7) uses:

1. Icons that are also meters (the heaters and the tank)
2. Icons that rotate (the valve before the turbine)
3. Trend plots with digital readouts
4. Trend plots with associated bar graphs

━━━━━━━━━━━━━━━━━━━━ ▲▼▲ ━━━━━━━━━━━━━━━━━━━━

Work Domain Analysis

A WDA was performed for the simulated reactor. Because the analysis is similar at many levels to the DURESS analysis and the thermal generating plant in Figures 6.2 and 6.3, we have not shown the analysis here.

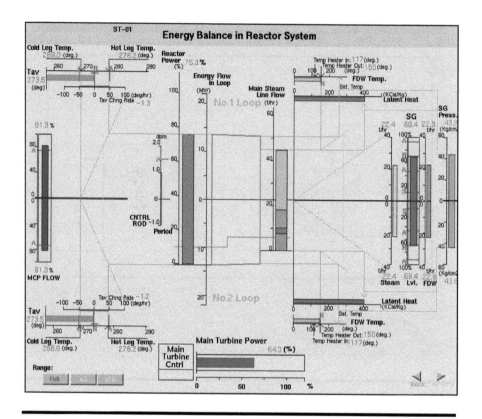

Figure 6.8 Display for "Energy Balance in Reactor System" in a normal condition. (Reprinted from Yamaguchi, Y. and Tanabe, F. (2000) Creation of the interface system for nuclear reactor operation — practical implication of implementing EID concept on a large complex system. *Proceedings of the XIVth Triennial Congress of the International Ergonomics Association and the 44th Annual Meeting of the Human Factors and Ergonomics Society, 3, 571–574.* © 2000 by the Human Factors and Ergonomics Society. All rights reserved.)

─────────▼▲▼─────────

EYE ON VISUALIZATION

This display (Figure 6.8) uses:

1. A rectangular configural feature (under No. 1 loop)
2. Triangular geometric relations shown by red dashed lines
3. Meters in different orientations
4. A stacked bar graph (center in blue)

─────────▲▼▲─────────

THE ANALYSIS IN THE DESIGN

Analysis: Similar to Figure 6.3

This display shows several higher WDA levels. The key features are:

FP:	Power demands	Main turbine power meter at the bottom of the display
AF:	Energy balances	Energy transfer between loops in central graphic
	Mass balance	Mass flow meter, central
GF:	Temperature and	Meters right hand side pressure
	Connecting to AF	Triangular lines connecting temperature and flow to energy

Display Design

Five displays representing higher level information were created: Energy Balance in Reactor System, Mass Balance in Primary System, P-T Diagram with the Functional Relation of Related Process Parameters, Mass Flow Distribution of Whole Secondary System, and State Diagram of Steam Generator. We have focused our discussion on the first display, the energy balance for the reactor. This display is shown in Figure 6.8.

Energy transfer for the reactor loop to the main steam line is shown by a rectangle in the center of Figure 6.8. The height of the left edge of the rectangle shows the energy in the reactor loop, and the height of the right edge of the rectangle shows the energy transferred to the other loop. Under normal conditions, these levels are balanced. In the abnormal situation shown in Figure 6.9, more energy is being produced by the reactor than is being transferred to the other loop.

The designers have also used the triangle relations to show mass and energy relationships visually. These graphics are influenced by the DURESS graphic discussed in Chapter 1. In this case it should be noted that the triangle sidelines in gray are often extended a fair distance across the display (visible in Figure 6.8 but truncated in Figure 6.9).

Evaluation

The displays developed for the reactor system were implemented in a full-scope simulator. Evaluations are still in progress; however, the following observations were noted from preliminary assessments:

1. The control performance particularly in normal operation was considerably improved.
2. The new information items at higher levels are very effective at providing overall situational context.

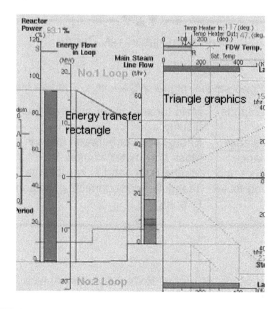

Figure 6.9 Key elements of the display in an abnormal condition. (Reprinted from Yamaguchi, Y. and Tanabe, F. (2000) Creation of the interface system for nuclear reactor operation — practical implication of implementing EID concept on a large complex system. *Proceedings of the XIVth Triennial Congress of the International Ergonomics Association and the 44th Annual Meeting of the Human Factors and Ergonomics Society,* **3, 571–574. © 2000 by the Human Factors and Ergonomics Society. All rights reserved.)**

DESIGN FOR A PASTEURIZER

Reising and Sanderson (2002a) improved upon an existing microworld simulation of a pasteurizer. They implemented EID displays for the microworld. As well, the microworld was configured for studying operator adaptation and, in particular, the impact of instrumentation configuration and reliability on highly configural displays. This work is included as an example of an EID, as well as for its insight into determining adequate sensor levels for EID.

The purpose of a pasteurizer is to heat milk from an initial input temperature to a final temperature within a particular range while maintaining a certain minimum flow rate. The milk must be held within the target temperature range for 15 seconds to be legally pasteurized (Hall and Trout 1968; cited in Reising and Sanderson 2002a). Figure 6.10 is a schematic of the simulated pasteurizer that was used in the Pasteurizer II project. The key aspects of the system to notice are the input line (top left corner), the water heater and hot water loop, and the product line in blue that

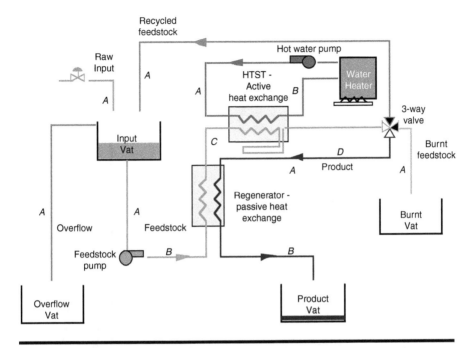

Figure 6.10 Pasteurizer II structure. (From Reising, D. V. C. (1999) The impact of instrumentation location and reliability on the performance of operators and ecological interface for process control. Unpublished thesis, University of Illinois, Champagne-Urbana.)

includes a heat regeneration phase. (At a certain level, this is a two-loop structure similar to the structure of Tanabe's nuclear reactor in Figure 6.8.)

System Boundary

The system boundary is as shown in Figure 6.10.

Work Domain Analysis

Reising and Sanderson conducted a WDA, developing an Abstraction Hierarchy, of the pasteurizer system shown in Figure 6.10; the analysis is shown in Figure 6.11. The WDA has four levels and does not represent Physical Form directly. However, the authors included many Physical Form attributes of the simulation in their eventual display. For example, the time lags in the simulation depend on pipe lengths, diameters, vessel sizes, and so on. The mimic display presents the piping in its proportionate lengths as simulated. So while Physical Form information is not directly presented, it is contained in the simulation and the eventual displays. For the other four levels, you should note the somewhat different analysis and representation from the other examples in this chapter.

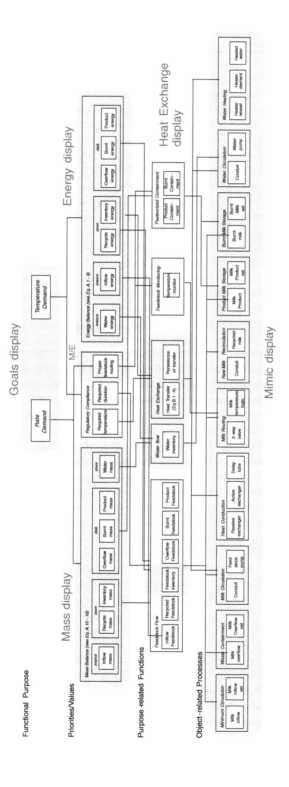

Figure 6.11 Work domain model for Pasteurizer II, showing plan for display design. (Reprinted with permission from Reising, D.C. and Sanderson, P. (2002). Ecological interface design for pasteurizer II: a process description of semantic mapping, *Human Factors*, 44(2). ©2002 by the Human Factors and Ergonomics Society. All rights reserved.)

First, some of the Part-Whole Decomposition is shown directly in the hierarchy. The authors have placed the part-whole elements as nodes within the more aggregate elements. This representation has the advantage of showing the connection between an aggregate element and its part-whole hierarchy very closely. The other difference you should note is that the abstraction dimension of the analysis differs slightly from the method that we have been following. The labels of the levels differ and the definition of the levels, particularly the lowest level, Object-Related Processes, is also different. This level is notably more process related in its definition than the Physical Function levels shown in previous models. Much of this information would have been placed at Generalized Function in the other models. The main features of the analysis are the same, though, meaning that the analysis decomposes the systems by both part-whole and means-end relations. The priority/value label for the second level may allow for a richer description when using the analysis beyond process control systems and the Object-Related Processes level shows a tight connection between objects and their end processes. There is some redundancy possible between this level and the third level (Purpose-Related Functions) since this level has an alternative, though more abstract, process description. We would not expect significantly different display results from these analytical differences, but we included this model to show alternatives in WDA.

Functional Purposes: The top level shows the Functional Purposes of the system: the Flow Rate Demand and the Temperature Demand.

Priorities/Values: This level is largely equivalent to Abstract Function. The analysis has modeled mass and energy balances as well as regulatory compliance. Regulatory compliance in this case relates to holding the milk at the required temperature for the required time. Arguably, this requirement could have been included at the Functional Purpose level as well.

Purpose-Related Functions: At the third level are functional processes such as water flow, milk flow, and heat exchange. These processes provide the fundamentals for the pasteurizer system.

Object-Related Processes: This fourth level is also process related (note the similarity of language in the third and fourth levels). At this level, though, the processes are more concrete and describe what is needed to attain the processes at the third level. For example, heat exchange is connected to milk flow, water flow, and heat conduction, which are all needed for effective heat exchange.

Sensors and Information Availability

An interesting aspect of Reising and Sanderson's work is that they explored the system under two different levels of information availability. The levels differed as shown in Table 6.1.

Table 6.1   Sensor Levels for Pasteurizer II

Level	Maximum Configuration	Minimum Configuration
Functional purpose	Rate and temperature sensed	Flow rate approximated Temperature sensed
Abstract function	Mass and energy flow rates derived, but derived from independent sensors	Some derived independently, most borrowed from other derivations
Generalized function	Volumes, flow rates, and temperatures sensed Heat exchange rates derived	Most derived or borrowed from other sensors
Physical function	36 sensors and 3 control settings	13 sensors and 3 control settings

Interface Design

Figure 6.12 shows the interface. The designers made extensive use of their analysis in Figure 6.11, developing a functional information profile to use in planning their interface. We have discussed their information profile, their use of graphical elements, and overall display organization in the sections below. Following this section, evaluation results are summarized.

━━━━━━━━━━━━━━━━━ ▼▲▼ ━━━━━━━━━━━━━━━━━

FOCUS ON TERMS

Display — A visual representation of the domain
Interface — A display that includes the ability to control the domain

━━━━━━━━━━━━━━━━━ ▲▼▲ ━━━━━━━━━━━━━━━━━

Functional Information Profile: The gray fields behind collections of nodes shown in Figure 6.11 indicate functions or properties from which a configural display was developed (Reising and Sanderson 2002). For example, there is a goals display, a mass display, and an energy display. The display plan focuses on the critical aspects of the system, largely the heat exchange, and the designers have planned certain displays to cover each level of the analysis. This shading is equivalent to developing a Functional Information Profile, as discussed in Chapter 4. The authors showed a complete Functional Information Profile in Figure 10 of their paper (Reising and Sanderson 2002).

Graphic Form Design: In the goals part of the display (top left corner), the demand and flow rates are shown against their targets using a straightforward single-bar indication. The mass display uses bar graphs in

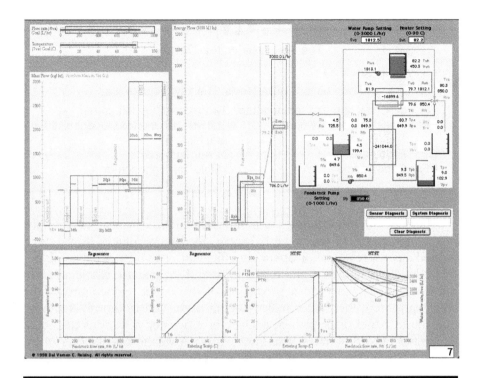

Figure 6.12   Display for Pasteurizer II. (Reprinted with permission from Reising, D.C. and Sanderson, P. (2002). Ecological interface design for pasteurizer II: a process description of semantic mapping, *Human Factors*, 44(2). ©2002 by the Human Factors and Ergonomics Society. All rights reserved.)

a configural layout, and the energy display takes advantage of parallel lines in a rectangular form to effect a configural feature, the rectangle for heat exchange, similar to the rectangle used by Tanabe. Although the displays are separated by abstraction level, the designers connected elements across the displays by using color coding.

▼▲▼

EYE ON VISUALIZATION

This display (Figure 6.12) uses:

1. Rectangular configural features
2. Horizontal meters in the top left corner as an overview
3. Variables plotted against a background of properties in the lower left graph
4. Careful use of color to manage salience, only values and thresholds are colored

▲▼▲

▬▬▬▬▬▬▼▲▼▬▬▬▬▬▬

THE ANALYSIS IN THE DESIGN

Analysis: Figure 6.11

This display shows all four WDA levels. The key features are:

FP: Milk flow rate and temperature   Two meters, top left corner
AF: Mass and energy flow   Transfer between loops central left graphics
GF: Processes   Lower four charts
PFn: Components   Mimic, top right corner

▬▬▬▬▬▬▲▼▲▬▬▬▬▬▬

There are several examples of multivariate graphic forms. For example, in the first panel of the lowest display, feedstock volume flow rate is plotted against regenerator operating efficiency. The next panel combines two inlet temperatures to the regenerator and regenerator efficiency from the first panel to provide a prediction of the outlet temperature of the cold leg (where the colder milk is being preheated in the regenerator). The fourth panel displays heat exchanger efficiency at various feedstock and water mass flow rates. The third panel allows the calculation of cold leg outlet temperature using the efficiency value from panel 4. Panels 2 and 3 exploit linear relations by mapping system equations to a straight line:

Panel 2 and 3 equations:

$$T_{co} = \varepsilon(T_{hi} - T_{ci}) + T_{ho} \text{ maps to } y = aX + b$$

Panel 4 exploits an exponential relationship generating the more complex series of curves where:

$$\varepsilon = 1 - e(f(M/M)/1 + (M/M)$$

Interface Organization: The interface is organized with Functional Purpose information in the top left. Moving to the right on the display, the levels of abstraction lower to the mimic display at the top right. This organization maps the flow of meaning, from goals to concrete equipment, to a natural information search flow or reading pattern of left to right. It is also typical to map the abstract to the concrete along the vertical dimension, with more abstract elements at the top and more concrete near the bottom. The designers have also chosen to keep all the information on a single display page.

Evaluation

Participants worked with the Pasteurizer II display for five sessions with either the ecological interface or a mimic-only interface with limited training or prior exposure to the system. The ecological interface supported better fault diagnosis, but only when supported with adequate instrumentation. In ratings of usefulness, participants reported that the energy display was important for reaching higher performance levels.

Recall as well that Reising and Sanderson evaluated their interface with two levels of sensor information. He found more correct diagnoses in the high-sensor condition, for both interfaces. Control performance was the same, regardless of interface or level of sensor information.

DESIGN FOR AN ACETYLENE HYDROGENATION REACTOR

This case involved the analysis and design of ecological displays for the Acetylene Hydrogenation Reactor, one subsystem used in an ethylene processing facility (Miller and Vincente 1998; Jamieson 2002). This reactor receives a partly processed mixture of ethane (C_2H_6), ethylene (C_2H_4), and acetylene (C_2H_2). The purpose of the reactor is to remove the acetylene by converting it to ethylene ($C_2H_2 \rightarrow C_2H_4$). This is done through a chemical reaction that adds hydrogen to the acetylene molecule. As a side product, some of the ethylene is converted to ethane in the same manner, although ethane is not a desired product. The purpose of the process is to maximize the removal of acetylene, maximize the production of ethylene, and minimize the production of ethane. The target concentration of acetylene out of the reactor is less than 5 ppm.

System Boundary

The boundary included the general acetylene reactor facility with its corresponding equipment. The designer opted to include a pyrolysis unit in the analysis as well. Although this unit was physically distant from the acetylene reactor unit, the pyrolysis unit produces hydrogen (H_2) and carbon monoxide (CO). Hydrogen is a required input for the reaction, and carbon monoxide is critical to controlling the hydrogenation reaction. The reactor operators typically were given some control over the pyrolysis furnaces, confirming that this unit belonged within the system boundary.

Work Domain Analysis

Part-Whole Hierarchy

Three levels of decomposition were used in the Part-Whole Hierarchy. The decomposition levels were reactor, units, and components. Figure 6.13 shows this hierarchy.

Abstraction Hierarchy

The Abstraction Hierarchy was conducted at five levels (Miller and Vincente 1998), with both functional and causal models of the work domain. In this

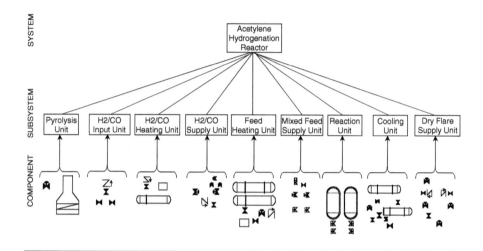

Figure 6.13 Part-Whole Decomposition of the acetylene reactor. (From Miller, C. A. and Vicente, K. J. (1998) Abstraction decomposition space analysis for NOVA's E1 acetylene hydrogenation reactor (CEL-98-09), Cognitive Engineering Laboratory, University of Toronto.)

case the causal models of the work domain have been shown. For a similar functional model, see the Pasteurizer example earlier in this chapter.

Functional Purpose: The main purpose of the reactor was to remove acetylene, reducing it below the maximum level of 5 ppm. The authors acknowledged the existence of other purposes (e.g., safety, profitability, environmental protection) but did not include them in the model at this point.

Abstract Function: The Abstract Function layer shows the mass and energy movements through the plant. Mass and energy flow were separated; Figure 6.14 shows mass flows, and Figure 6.15 shows energy flows.

The designer chose a second modeling technique, Multilevel Flow Modeling (MFM), to express these levels. Multilevel Flow Modeling is a functional modeling approach that is well suited for expressing functional flows of mass or energy. Multilevel Flow Modeling defines functional types or primitives of source, sink, transfer, barrier, and transport. Each primitive has a specific symbology associated with it; for example, sources are circles with a black dot in them, sinks are circles with an x through them. For more information on Multilevel Flow Modeling, see Lind (1990; 1991; 1994).

Generalized Function: The Generalized Function layer modeled the chemical reactions that occurred in the reactor as well as the required flows, heating, cooling, and pressure changes. Figure 6.16 shows the Generalized Function level of the model. Note the complexity of flows in the system.

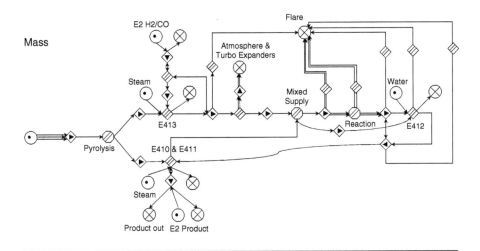

Figure 6.14   Mass flows through the reactor. (From Miller, C. A. and Vicente, K. J. (1998) Abstraction decomposition space analysis for NOVA's E1 acetylene hydrogenation reactor (CEL-98-09), Cognitive Engineering Laboratory, University of Toronto.)

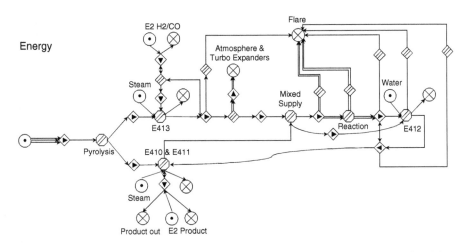

Figure 6.15   Energy flows through the reactor. (From Miller, C. A. and Vicente, K. J. (1998) Abstraction decomposition space analysis for NOVA's E1 acetylene hydrogenation reactor (CEL-98-09), Cognitive Engineering Laboratory, University of Toronto.)

Physical Function: This level showed the components and considered their settings and capabilities. Figure 6.17 shows the Physical Function level of the model. Note that the causal depiction of this level is very similar to a piping and instrumentation drawing of the plant.

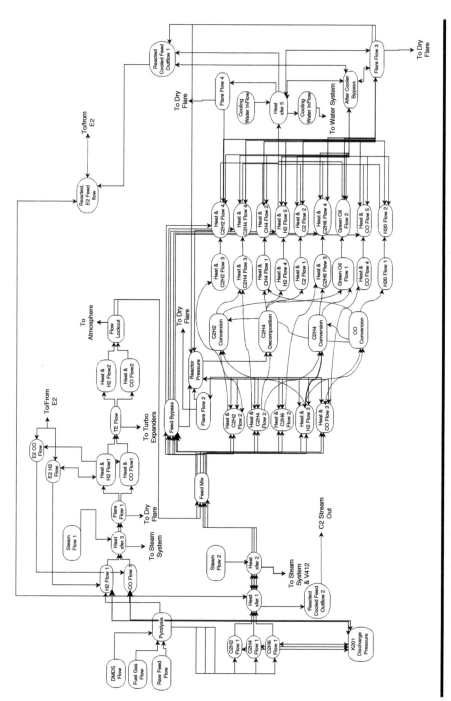

Figure 6.16 Generalized function level, acetylene reactor. (From Miller, C. A. and Vicente, K. J. (1998) Abstraction decomposition space analysis for NOVA's E1 acetylene hydrogenation reactor (CEL-98-09), Cognitive Engineering Laboratory, University of Toronto.)

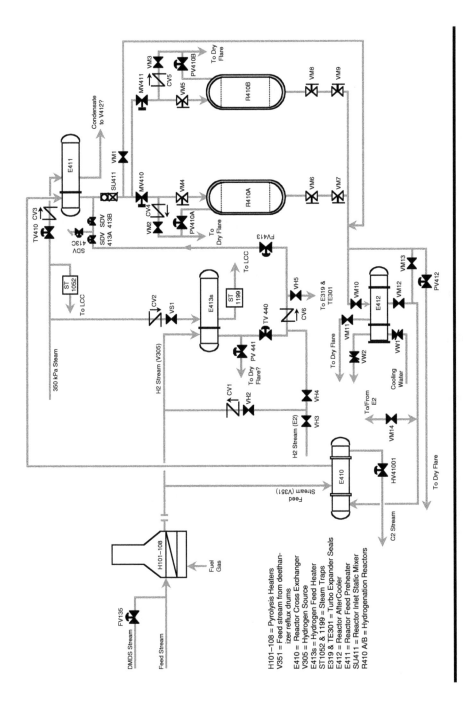

Figure 6.17 Physical Function for the acetylene reactor. (From Miller, C. A. and Vicente, K. J. (1998) Abstraction decomposition space analysis for NOVA's E1 acetylene hydrogenation reactor (CEL-98-09), Cognitive Engineering Laboratory, University of Toronto.)

Physical Form: The Physical Form layer concentrated on the appearance and location of the equipment in the plant.

Task Analysis

Miller and Vicente (1998; 2001) performed a Hierarchical Task Analysis of this reactor. The results of this task analysis were used by Jamieson (2002) to create an interface that contained both WDA and task analysis information. Figure 6.18 shows a small part of the task analysis. Note how, in contrast to the work domain model, the task analysis shows operational steps, the order of steps, and the conditions for changing steps. This represents a different, but complementary, set of information requirements. As Jamieson's work revealed, having both WDA and task analysis information can improve operation. We have included this example because it is a very clear example of using the two methods in an effective fashion. For more discussion on comparing WDA and other methods, and how to fit these together in a complete design process, see Chapter 10.

Interface Design

In creating this design, the designers performed three different analyses. They did a WDA, a Hierarchical Task Analysis, and a Control Task Analysis. Then they looked at the information specified by each of the analyses. Table 6.2 gives an example of the different information requirements from the different analyses for a part of the design — that of a ratio controller. This table is adapted from Jamieson (2003b), and Jamieson (2003a) contains a richer analysis of information requirements. In particular, we have shaded in gray those information requirements that did not come from the work domain model. In particular, note that the WDA revealed components, flow, and relationships. The Control Task Analysis in particular identified setpoints, current values, and important error measures. Some of this information would also have been available by looking at variable constraints, as suggested in Chapter 4.

The designers developed two different displays, an EID display that used the information from the work domain model only, and an EID+Task display that used the information from the work domain models, Hierarchical Task Analysis, and Control Task Analysis. Figure 6.19 shows the EID+Task display.

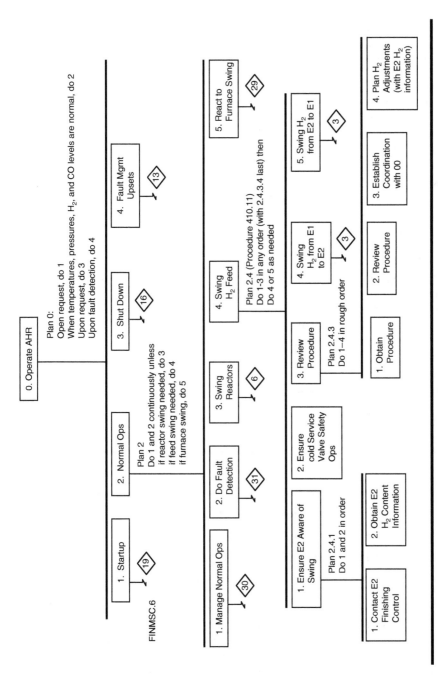

Figure 6.18 Part of the Hierarchical Task Analysis of the acetylene reactor. (From Miller, C. and Vicente, K. J. (1999) Comparative Analysis of Display Requirements Generated via a Task-Based and Work Domain-Based Analyses in a Real World Domain: NOVA's Acetylene Hydrogenation Reactor (CEL-99-04), Cognitive Engineering Laboratory, University of Toronto.)

Table 6.2   Information Requirements for the Acetylene Reactor and Their Sources (HTA Refers to the Hierarchical Task Analysis; CTA Refers to the Control Task Analysis)

Information Requirement	WDA	HTA	CTA
Flow valve (appearance and location)	α	α	α
Flow valve setting	α	α	α
The effect of flow valve setting on flow through the valve	α	α	α
The flow of CO into the reactor	α	α	
The flow of H_2 into the reactor	α	α	
The flow of C_2H_2 into the reactor	α	α	
The flow of C_2H_4 into the reactor	α	α	
The flow of C_2H_6 into the reactor	α	α	
The effect of the setting of a valve on the flow of H_2	α		α
The effect of the state of a pressure indicator on the flow of H_2/CO through a temperature sensor	α		
The effect of the setting of temperature indicator on flow of H_2/CO	α	α	α
H_2/CO heat and supply input	α		
C_2 heat and supply input	α		
H_2/C_2 weight ratio setpoint		α	α
H_2/C_2 weight ratio present value		α	α
H_2/C_2 weight ratio error (setpoint: present value)			α
H_2/C_2 weight ratio valve percentage open			α
FC setpoint			α
FC percentage open			α
Error (FC setpoint: operating point)			α
Differential A status			α
Temperature operating point			α
Low flow limit on present value of 40 Mg/hr		α	

▼▲▼

THE ANALYSIS IN THE DESIGN

Analysis: Figures 6.13 through 6.18, Table 6.2

This display shows all four WDA levels. The key features are:

FP:	Remove acetylene	Change in acetylene graphic, acetylene meter, right hand side
AF:	Mass and energy flows	Dual bar graph, lower right corner
GF:	Processes for AC removal	Temperature profile in the reactor, temperature profile across components (lower center), H_2/CO/AC graphics
PFn:	Components	Settings, lower half of the display

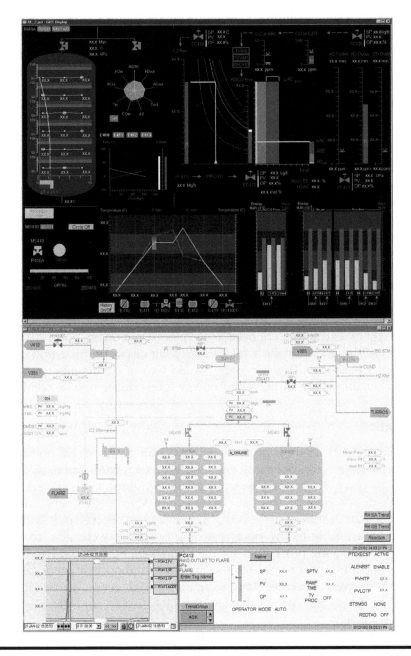

Figure 6.19   Display with WDA and task analysis information for the acetylene reactor. (Reprinted with permission from Jamieson, G. A. (2002) Empirical evaluation of an industrial application of Ecological Interface Design, *Proceedings of the 46th Annual Meeting of the Human Factors and Ergonomics Society*, 536-540. ©2002 by the Human Factors and Ergonomics Society. All rights reserved.)

In the display, the designer separated physical mimic-based information (on the lower half of the screen) and functional, higher level information (on the top half of the screen). The functional part of the display contains several graphical elements that we have discussed before. In the top left corner of the display is a highly integrated reactor graphic showing temperatures at different locations in the reactor. The temperatures are shown using a series of temperature graphs with the measures connected by a vertical line. The vertical line reveals the temperature profile throughout the reactor and serves as an emergent feature, allowing the operators to diagnose a number of conditions. To the immediate right of this display is a polar star display with nine variables. To the right of this is the graphic for ratio control, showing the variables from Table 6.2.

In the second row of the display, a temperature profile is given, showing temperature gains and losses through the reactor and the heat exchangers before and after the reactor. The operators would be looking for an expected pattern of temperature increases through the heat exchangers before the reactor and temperature decreases through the exchangers after the reactor. (Note how some exchangers are used before and after the reactor to exchange heat from the reacted products in order to heat the reactor feed. This type of exchanger connection is quite common in process control plants.) At the bottom right-hand corner of the display, a mass and energy bar graph is shown. This bar graph would have a configural aspect to it, with a certain pattern of mass and energy levels under normal conditions.

Evaluation

Jamieson (2002) used 30 professional operators in his evaluation of the interface. Their mean operating experience was 3.6 years and ranged from 0.5 to 10 years. The display was evaluated with three kinds of events: an "Anticipated-Normal" event, an "Anticipated-Abnormal" event, and an "Unanticipated-Abnormal" event. These events correlated with the resources available to the operators; i.e., operators were able to use a "standard operating procedure" for the normal event, they used an "emergency operating procedure" for the anticipated abnormal event, and no procedures were available for the unanticipated event. Three displays were evaluated: the current display, the EID display, and the EID+Task analysis display.

Trial completion times were fastest in the EID+Task analysis display, followed by the EID display, and finally the current display, with statistically significant differences between the current display and the EID+Task display. With both the EID and EID+Task displays, operators took fewer control actions than with the current display. Operators made more correct fault diagnoses with the EID+Task display than with the other two displays.

Alternative: Design for a Fluid Catalytic Cracking Unit

The same designer developed a display for a fluid catalytic cracking unit (Jamieson and Vincente 2001). We have included the display to show how various graphical elements have been used. Figure 6.20 shows the display. We have used the annotations to emphasize the graphical elements present in the display. The annotations are above the graphical element. The designers have used polar star displays to integrate the turbine variables. In several locations, triangle graphics, inspired by the DURESS graphic, are used to show mass and energy relations. There are extensive uses of variable balances, and a Mass Data Display is used as an overview. We have not included the analysis behind this display; it has similarities to the analysis for the acetylene reactor. The display has not been tested.

DESIGN FOR A LARGE REFINERY

The Syncrude extraction and upgrading facility in Ft. McMurray, AB, Canada, is one of the largest oil refining facilities in the world and one of the first to implement EID in its plant. The Syncrude plant takes oil embedded in sand, extracts this oil, and processes it into lighter oil products. The plant itself is divided into individual plants, each charged with a certain phase of the overall process, be it extraction, oil refining, removal of noxious environmental elements, or production of process requirements such as steam.

The analysis phase of this project involved seven analysts over the course of a year. EID concepts and drawings were delivered to the control system providers for implementation. The project is still under way at the time of writing this book; therefore, only design concepts are shown and no evaluation results are available yet. The project is included here as an example of a large-scale implementation of EID and because it offers insight into managing a large EID project.

System Boundary

The system was bounded to include the contents of the plant up to the metering station, where product was sent down the pipeline. Inputs from mining operations were not included, and some auxiliary systems (e.g., steam lines and venting) were not intensively modeled.

Work Domain Analysis

Part-Whole Analysis

This project began with an extensive and explicit Part-Whole Analysis over a number of months. Analysts took different plants and broke the plants down according to the levels shown in Figure 6.21.

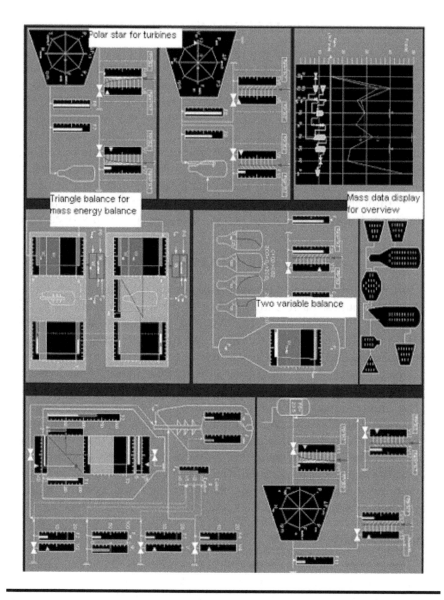

Figure 6.20 Display for a fluid catalytic cracking unit. (From Jamieson, G. A. and Vicente, K. J. (2001) Ecological interface design for petrochemical applications: Supporting operator adaptation, continuous learning, and distributed, collaborative work, *Computers and Chemical Engineering*, 25: 1055-1074.)

Important aspects of this analysis were that the decomposition levels were planned before beginning the analysis. It was decided to end the analysis before the component level. In this case, component level drawings

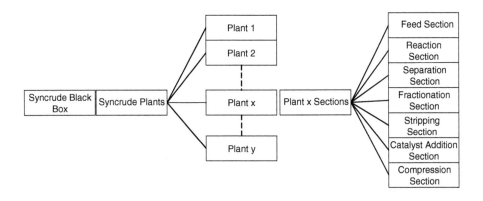

Figure 6.21 Part-whole hierarchy plan for the Syncrude project.

were available from plant drawings. Given the complexity of plants, there was little value in redeveloping this level. At each level in Figure 6.21, a separate Part-Whole Decomposition was performed, resulting in hundreds of individual Part-Whole Decompositions.

Abstraction Hierarchy for a Reactor

At each level of Part-Whole Decomposition, an Abstraction Hierarchy was created. In its entirety, the collection of part-whole drawings and Abstraction Hierarchy drawings comprised the work domain model. Figure 6.22 shows an Abstraction Hierarchy for a reactor in one of the plants. (Specific plant details have been removed.) The traditional Abstraction Hierarchy labels were changed to Purposes, Principles, Processes, and Components, in order to improve communication with plant engineers.

There are several important aspects to notice about the analysis. First, at the level of Purposes, there were production purposes (such as convert bitumen to product), product quality purposes (such as maintain constant reactor temperature profile at target temperature and pressure), and economic purposes (such as maintain catalyst level and density). In many parts of the analysis there were also environmental purposes (such as reduce emissions or reduce sulfur content of the product). At the Principles level, mass and energy structures were modeled separately. This was because the mass and energy flows had distinctly different structures due to the heating and cooling of product, and exothermic reactions. Processes and Components are modeled. Figure 6.22 shows a mostly causal model with functional means-end links.

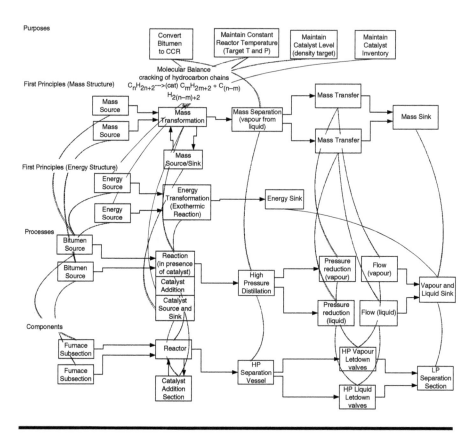

Figure 6.22 One part of the Syncrude work domain model, reactor level.

Interface Design

The following figures show ecological design concepts, rather than actual screen shots (the displays were undergoing implementation at the time of writing this book). Figure 6.23 shows an overview concept for the reactor. Reactor temperatures are normalized and form the vertical line under normal conditions. This represents the Functional Purpose objective of maintaining constant reactor temperature. The bar graph in the reactor reports the level catalyst from density detectors, for maintaining catalyst level and inventory. Below the reactor are two small graphs showing the relationships between reactor temperature and feed rate on product conversion, and catalyst addition rate on product quality and desulfurization. These graphs take Generalized Function variables (e.g., reactor temperature, feed rate, catalyst addition rate) and connect them to Functional Purpose level requirements (pitch conversion and desulfurization). The shaded gray areas show the regions of optimal and most economical

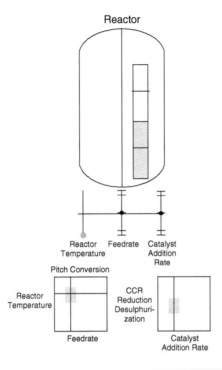

Figure 6.23   Overview screen concept, Syncrude project.

performance for the reactor as determined by plant engineers. A balance graph made of three indicators is shown in the middle of the graphic and depicts the two variables that can influence reactor temperature, feed rate, and catalyst addition rate. This graphic provides integration of data from the square graphs below and connects to the reactor temperature line above. This is an example of an overview graphic that would be supported by more detailed plant information on another screen.

HANDLING THE CHALLENGES

In the various examples in this chapter, we've seen analyses for relatively larger scale systems than in other parts of this book. As a result, the challenges are different from the other examples. Reising and Sanderson's work on the Pasteurizer begins to address the challenge of matching sensor levels to ecological displays. Also in this chapter, the various designers have employed different variations of the basic WDA to handle these systems. The Syncrude example provided an example of how to organize an extremely large analysis project. In the Pasteurizer project we see a functional analysis with some of the Part-Whole Analysis included in it. In the acetylene reactor we saw a distinct part-whole phase, then causal models for each of the levels. In this project, the causal models included another modeling approach: Multilevel Flow Models.

These decisions do not affect the fundamentals of the modeling exercise or the information obtained from the models, but they can make a big difference in terms of organizing that model, describing the model to other people, and developing a design from the model. We will look at the differences we see here in terms of two dimensions: working with the Part-Whole Analysis, and then modeling flows and functionality.

Our final challenging area is the organization of information in these large displays. Ecological information is unique in that it provides several different layers of information for the same work domain. These layers have to be managed effectively so that the user establishes effective patterns of information search, rather than being overwhelmed by the additional information. This chapter gave several different examples of ways of organizing displays.

Challenge 1: Instrumentation or Sensor Availability

Reising and Sanderson's work on Pasteurizer II makes the strongest statement on the impact of sensor availability on ecological interfaces. They show that a full sensor set does improve diagnostic performance with ecological interfaces. Furthermore, implicitly their work points to how WDA can be used as a tool to determine sensors and their placement.

It may be useful to indicate directly sensed versus derived information in your interface. No one has yet studied the effects of sensor noise or information uncertainty on ecological interfaces.

Challenge 2: Part-Whole Analysis Variations with Project Size

Three alternatives were shown here, part-whole integrated with abstract, one explicit Part-Whole Hierarchy, and an extensive set of Part-Whole Analyses. The alternatives seen here are largely variations on the scale of the project. For example, in the Pasteurizer project, the part-whole dimension is integrated with the abstract dimension. The designers used a functional representation and then, within that representation, included a part-whole breakdown at certain points. The system being analyzed has 11 main components, and it is possible for the designers to show both the functional and part-whole breakdowns in the same figure (Figure 6.11). The inclusion of the part-whole dimension shows the tight connections between the levels and saves the analyst from having to produce a second level of model.

Table 6.3   Part-Whole Representation and Project Size

Size of Project	Effort on Part-Whole	Representation	Example
0–20 components	Implicit analysis	Can include with abstract	Pasteurizer
20–1000 components	Explicit analysis	Single hierarchy or work domain space	Thermal, acetylene reactor
1000 + components	Explicit analysis	Multiple hierarchies, used for organization	Syncrude

The second approach to treating the Part-Whole Analysis is demonstrated in the acetylene reactor and, to a lesser extent, the thermal generating station. Both of these projects have in the range of 200 to 500 distinct variables, thereby representing medium-sized implementations. In these cases, the designers did a single-model Part-Whole Analysis at the levels of system, subsystem, and component, then generated abstraction models at each of these levels. The acetylene reactor showed the part-whole dimension as a hierarchy (Figure 6.13), whereas the thermal generating plant showed the Part-Whole Hierarchy as an abstraction-decomposition space (Figure 6.2). Both systems would have been too complex to use the tight representation shown in the Pasteurizer project.

The third approach, from the Syncrude project, is really a larger scale example of the second approach. In this case, multiple part-whole models are explicitly developed and detailed at several levels: overall plant, individual plants, plant sections, and individual sections. The system in this case includes thousands of components. The analysis was being conducted across seven analysts, and maintaining consistency was a concern. At each of the part-whole levels, a full abstraction model is created. The Part-Whole Analysis in this project was a key organizing tool for the other analyses.

From these observations, we suggest the approaches shown in Table 6.3.

Challenge 3: Using Functional, Causal, and Multilevel Flow Models

In detailing their Abstraction Hierarchies, several approaches were shown in this chapter. The Pasteurizer example showed a functional Abstraction Hierarchy; the acetylene reactor and Syncrude projects showed causal abstraction models; and the acetylene reactor project used Multilevel Flow Models at the level of Abstract Function. Though not shown, the Pasteurizer project also used causal models, and the thermal generating plant project also used causal models and Multilevel Flow Models.

Table 6.4   Choosing Functional, Causal, and Multilevel Flow Models

Choice Factors	Types of Models	Example
Smaller system, straightforward flows, mass and energy flows are similar or not complex	Functional only	
Larger systems, elaborate flows, mass and energy flows are similar or not complex	Functional + Causal	Pasteurizer, Syncrude
Complicated systems where mass and energy flows may be different	Functional + Causal + MFM	Thermal generating system, acetylene reactor

Functional models portray means-end links most effectively. The connections here are direct and illustrated for people using the analysis. In many cases, though, the analyst needs to follow the flow structure as well to understand how the system works. This is when you would want to start using causal models. When the system is smaller, and the flows can be derived from the system diagram fairly directly, and the mass and energy flows are not complex, it may be possible to only do a functional model for your project.

As flow patterns become more complex, it becomes useful to add causal models to the functional models. This is seen in the acetylene reactor project and the Syncrude project. The causal models help to detail the flow structure and understand more complex flow patterns. In very elaborated flow systems, such as the Syncrude project, causal models can be used to a certain degree to simplify or abstract the flows. By this we mean that the analyst may find it useful to identify main feed and product lines at first, then control lines, emergency supply lines, or emergency shunting lines.

There are some cases where mass and energy flows are particularly complex or where mass and energy flows are distinctly different from each other. These are the cases where Multilevel Flow Models become useful. The process of understanding the functional primitives within the system and mapping the mass and energy flows separately can provide more detail. Systems with heat transfer or heat exchange are examples of systems where the mass and energy flow patterns are different from each other. The acetylene reactor project provides an example of using these models in the Abstraction Hierarchy. These models were also used in the thermal generating plant.

Table 6.5   Alternatives for Information Organization

Design Alternative	Examples
Top to bottom, left to right flow	Pasteurizer, thermal generating system
Overview and control screens	Acetylene reactor, Syncrude
Color coding	Pasteurizer
Shape consistency	Acetylene reactor
Spatial proximity	Thermal generating system

Challenge 4: Organizing Information on the Display

There are several alternative ways of organizing information on the display. The two main principles are the following:

1. You need to make the means-end links apparent, so that people can move smoothly between abstraction levels.
2. You want to encourage monitoring at the higher abstraction levels, with the more concrete levels reserved for control actions.

In this chapter, we saw three different approaches to organizing information. All three approaches, though, follow this principle, although in different ways. The nature of the design problem can constrain how this is implemented.

The Pasteurizer display organized the information so that the most abstract information was in the top left corner of the display. The designer then maintained as long as possible a left-to-right, top-to-bottom flow from abstract to concrete. This should encourage monitoring to start with the top left corner, or Functional Purpose information. To connect the means-end links more securely, they used color coding to reinforce the mapping between these display elements.

The thermal generating plant designers also used a top-to-bottom, left-to-right flow in their navigation tool. The particular features of the displays allowed the designers to connect means-end elements by spatial proximity, which under evaluation was an effective approach.

The Syncrude displays and the acetylene reactor displays utilized an approach where physical information is available on one screen, and functional information is available on one or more other screens. The objective behind this approach is to provide a monitoring or overview display that shows the functional information. Operators then transition to the screen with the physical information in order to get system details and take control actions. This is an effective design in situations where the system is so complex that the physical or mimic screens cannot portray additional information, and the functional information should remain visible for monitoring at all times. In designing these kinds of displays for a control room, you would generally aim to have the overview screen above or to the left of the physical information screen. The means-end connections between the two screens can be enhanced by color coding, layout, and shape consistency. The overview screen can also be used as a navigation tool.

Challenge 5: Adding Task Analysis Information to an EID

In Chapter 4 we showed that performing a WDA is rarely sufficient to design an effective display. The design needs more information on limits, thresholds, and key operating points in order to be useful. Jamieson's research with the EID+Task display shows this clearly. There are different ways of obtaining this information. You can perform Hierarchical or Cognitive Task Analyses, perform Control Task Analyses (more detail on this is available in Vicente 1999), supplement your work domain model with information from operating procedures, or use your work domain model as a guide for discussion with operators, following the questions posed in Chapter 4.

We also recommend that WDA be integrated into existing design methodologies. In Chapter 10 we discuss several methods and demonstrate how they can be used to generate different kinds of information in relation to EID. We use the same example throughout the chapter to make the demonstration more clear.

7

TELECOMMUNICATIONS
SYSTEMS

Telecommunications systems are primarily involved with sending information back and forth. They may be interested in connectivity between a sender and a receiver, access to applications, information throughput, or information security. These systems are more than just physical components; they also have software components in them. Mass and energy models do not fit these systems as well, since the modeling needs to capture the information flow of the system. These systems scale almost infinitely and are made of many, relatively similar components. Determining the system boundary is a challenge. In many ways, these systems are digital environments and have many unique features.

In this chapter we will discuss some of the unique challenges of modeling telecommunications systems. You will see three examples; two from network management and one from radio communication. These examples show very different work domain models from the other examples, since there is little emphasis on mass and energy transfer. You will also see a unique way of using the Part-Whole Decomposition, showing how the basic principles of the analysis can be maintained while the analysis is tailored to the needs of these very different systems.

CHALLENGES WITH TELECOMMUNICATIONS SYSTEMS

In our analysis of telecommunications systems, the basic principle behind the analysis is to capture the flow of information. In the radio system, this is the flow from sender to receiver. In the network situation, this is the flow from multiple points to multiple points. This flow of information, through electric signals, is the basis of these systems and therefore must be represented in the models.

▼▲▼

CHALLENGES WITH TELECOMMUNICATIONS SYSTEMS

Determining the system boundary
Determining WDA content
Developing a diagnostic graphic for a low-capacity situation
Using 3D ecological displays
Developing visualizations that scale with large amounts of data

▲▼▲

Determining the scale of the analysis can be difficult, particularly in networks. Networks connect to networks, and parsing at a network boundary is difficult. In the same way, developing a Part-Whole Decomposition is difficult, since in many ways, a Part-Whole Decomposition views the system at many different internal boundary lines. Furthermore, unlike process control systems, where each level of detail adds new information, networks look similar at all levels, just larger. To develop a more useful description, the network management example shows how the Part-Whole Decomposition can be transformed. In this example, instead of using physical device aggregates to define the dimension, information aggregates were used, which resulted in a more useful analysis. This presented one way of containing the scale of the network and developing a model that could accommodate various network sizes.

ANALYSIS FOR NETWORK MANAGEMENT

The analysis presented here was performed for a project with Nortel Networks, using a portion of the University of Waterloo Ethernet as a sample domain. From the work domain model, a display was developed and tested against a currently used network management display.

Network management is a surprisingly complex task. Most networks are complicated connections of old and new equipment and managed through an equally complicated set of tools. Big-picture views of the network world are rare, and truly diagnostic graphics are just as rare. Slow response times, networks that are down, and security threats like denial of service attacks result in real and significant costs. The diversity of networking equipment and the unpredictable creativity of people who hack into networks means that automated solutions are not likely to ever be completely successful.

System Boundary

In this analysis, we used a portion of our university's campus network (in particular, the University of Waterloo Village 1 residence network) as our network model upon which to demonstrate the principles of EID in the network management domain. This network provides Internet connections for 1,313 residence rooms in the Village 1 residence complex; it is composed of about 65 Ethernet switches that are centrally tied to the main switch and router module, which in turn connects to the main backbone switches on campus.

Determining the boundary for a network management problem, though, is quite difficult. Clearly, although we limited our analysis to this subset of the campus network, our devices are connected to the overall campus network and to the Internet in general. We, in effect, had to slice the analysis at the campus main switch to develop a usable system boundary.

Work Domain Analysis

As discussed in the previous section, determining a boundary for a computer network can be a challenging task. In a sense, networks have near-infinite connectivity, and there is not the easy division into subsystems that we see in other types of systems. This makes the determination of the Part-Whole Hierarchy difficult.

A first attempt at defining the Part-Whole Hierarchy along conventional divisions of network, subnet, and device was generally unsuccessful and revealed very little differentiation in terms of the constraints that were captured. We decided, therefore, to adapt the Part-Whole Decomposition and to map it onto the levels of the communication layer hierarchy. The communication layer hierarchy, instead of reflecting decompositions of devices, is a decomposition of communication layers. This resulted in our Part-Whole Hierarchy containing levels of the Application Layer, the Network Layer, the Data Link Layer, and finally, the Physical Layer. The last three layers map directly to the corresponding layers of the OSI Reference Model, while the top layer encompasses the Transport, Session, Presentation, and Application layers of the OSI model and is simply referred to as the Application Layer in this analysis. This was done because the top four layers of the OSI model all deal with end-user communication, and unlike the bottom three layers, the protocols used at the upper layers can vary significantly depending on the type of network used (Held 1998). Thus, to simplify the analysis, all four layers were grouped into one layer to represent application performance at the end-user level.

Figure 7.1 shows these dimensions and our work domain model at a fairly low level of detail. The following sections describe the contents of the model in more detail.

Functional Purpose

The purpose of a computer network is to provide a reliable and efficient means of communicating information between distributed applications

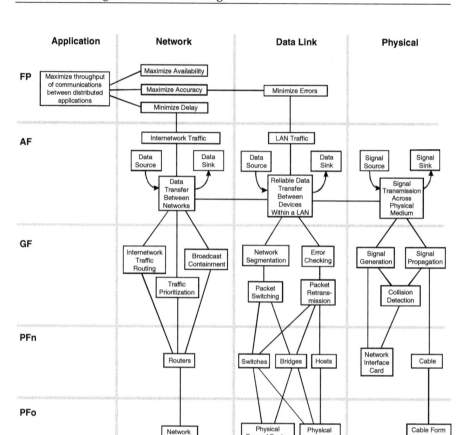

Figure 7.1   Work domain model for network management. (From Kuo and Burns 2000. ©IEEE.)

and devices. More specific purposes of a network are to maximize availability of devices, provide accurate information flow, and minimize delays.

Abstract Function

The Abstract Function (AF) level provides a view of the causal structure of the system. At this level, the network was described in terms of information flow from one point to another. Information is generated at some source, representing the originator or sender of the information, and is transferred through some medium to the intended recipient, or sink, for that information. The transfer mechanism must ensure that there are available paths to connect the source and sinks of the information flow.

The implication of this design is that information is preserved or "conserved" as it travels through the communication medium. While this conservation principle has a different meaning than that applied to mass and energy flow systems in most physical systems, it is nevertheless useful to describe communication networks in these terms. For example, information that is generated by a certain device (or source) and specifically intended for another device (or sink) must be transferred through the network to its destination without loss or modification to the original data. If the information from this source was somehow changed or misdirected to a host other than the intended recipient, then the conservation of information principle would be violated, since a loss of information has occurred between the source and sink. This concept was used at the different levels of decomposition.

Generalized Function

At the Generalized Function (GF) level, the representation of the work domain changes to a view of the processes by which the Abstract Functions are carried out. For example, at the Network Layer, the function of moving information through an interconnection of pathways from source to sink is achieved through the traffic routing process of the underlying system. At the Data Link Layer, the view of the domain is focused on the processes that support the transfer of data from device to device in a LAN: segmentation, framing, and switching. One of the primary measures at this level is transmission delay. Error checking is another process that is required at the Data Link Layer in order to satisfy the Functional Purpose of minimizing errors and ensuring accuracy of data transmission. This is done through a variety of error-detection protocols.

At the Physical Layer, the view of the network changes to the processes that occur at the signal level. As a result, most of the processes at this level are concerned with correct signal generation and propagation, as well as collision detection.

Physical Function

The Physical Function (PFn) level deals with the actual components and physical implementations of the processes described above. These include the actual physical routers, switches, bridges, hubs and cables that comprise the network infrastructure. The objects described at this layer of the Abstraction Hierarchy tell how the general network processes in the layer above are actually implemented.

Physical Form

At the Physical Form (PFo) level, components in the domain are described with respect to their appearance, location, and physical condition. This will include, for instance, the network topology and the relative geographic locations of devices that comprise the network. Such information is extremely important to a network manager who, for example, will need to quickly locate failed network components and perhaps visually inspect them to determine whether they need to be replaced based on their physical appearance or condition. In addition, network managers may need to quickly assess the physical capacity of individual equipment, such as the number of available ports in a switch, to determine whether or not excess equipment or other configurations need to be implemented when traffic patterns or the number of users and hosts changes.

Deriving Information Requirements from the Analysis

From the work domain model, information requirements were derived and described in terms of preliminary display requirements (Table 7.1). An instrumentation availability analysis was not conducted.

Table 7.1   Information Requirements Derived from the Work Domain Model

Abstraction Level	Information Requirements
Functional purpose	Performance target view
	Latency or overall response times
	Accuracy of information transmitted (usually not a problem)
	Availability of network resources
	Throughput of application data through the network
	Threshold settings
Abstract function	Information flow view
	Path availability between different LANs
	Traffic load and capacity on each LAN segment
	Balance of information flow within and between different LANs
	Resource utilization of individual links and devices
Generalized function	Routing and switching process view
	Traffic routing information
	Broadcast containment areas
	Rate of collisions and errors in different LAN segments
Physical function	Equipment view
	Information on capabilities, capacities, settings, and configuration of network devices
Physical form	Topological view
	Physical location of equipment
	Logical connections between network devices

Display Design

Following the analysis and derivation of information requirements, the next step was to use those requirements to design visualizations of the network. We proceeded systematically through each level, ensuring that the variables from that level were displayed. For graphical visualizations of the variables, we adapted some previously tested and successful visual techniques from other domains. A working prototype display was implemented in Visual C++ using the OpenGL library to render objects in a three-dimensional (3D) environment. The resulting application was a fully interactive environment, in which users can navigate to different parts of a network, select different objects of interest, and analyze resulting graphs and information provided. Thus, this prototype tool allowed us to demonstrate the effectiveness of the graphical concepts and ideas in a dynamic way. Since the purpose of the prototype was mainly to test the display design ideas and not to create a fully functional network management tool, the actual data used to drive the display were simulated from a collected network data.

The following sections provide a description of the prototype display at each level of the Abstraction Hierarchy used in the analysis.

Functional Purpose Display

The Functional Purpose display is an overview display that shows system performance against various objectives. This overview of system state remains continuously visible and allows the network manager to quickly identify network failures in any region of the network. The overview display is a dedicated view in the top left-hand corner of the display (shown in Figure 7.2). This view presents a 2D image of the topological layout of all the switches and routers in the network. Each device appears symbolically as a colored dot on the overview map, with the colors corresponding to the condition or state of the device at any moment in time (green for normal, yellow for warning, red for critical).

This display, with its multitude of dots, uses a large amount of individual data to show overall system state and is an implementation of what is called a "Mass Data Display" (MDD; Zinser and Frischenschlager 1994). MDDs have been shown to be very effective at supporting rapid fault detection and capitalizing on human pattern recognition skills. A trending bar graph is shown below the display (in Figure 7.2) to show network performance over time. The bar graph is stacked such that its height is determined by the sum of the number of devices that are currently in the "Warning," "Critical," and "Down " states. Consequently, a network operator can quickly tell how healthy the state of the network is or has been by observing the height of the bar graph. The taller the bars of the graph,

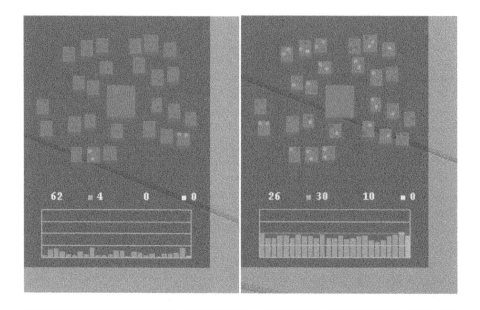

Figure 7.2 Two close-up shots of the overview map. The left shot is that of a relatively healthy network, while the one to the right depicts a less healthy one. (From Burns et al. (2003) Computer Networks, 43: 369-388. ©2003 Elsevier.)

the less healthy state the network is currently in, thus alerting the network manager that there may be potential problems occurring at the moment.

▼▲▼

EYE ON VISUALIZATION

This display (Figure 7.2) uses:

1. A mass data display with colored circles as the symbol
2. A digital count of summary statistics from the mass data display
3. A trending bar graph where the bars change color (from yellow to red) to show thresholds

▲▼▲

The overview map provides more than just monitoring purposes. It was also designed to serve as a navigation aid for the main 3D view (Figure 7.3). By positioning the mouse pointer and selecting a point on the overview map, the main view will immediately jump to the corresponding area of the network. In addition, the selected area is highlighted on the overview map, thus providing orientation information to the network manager.

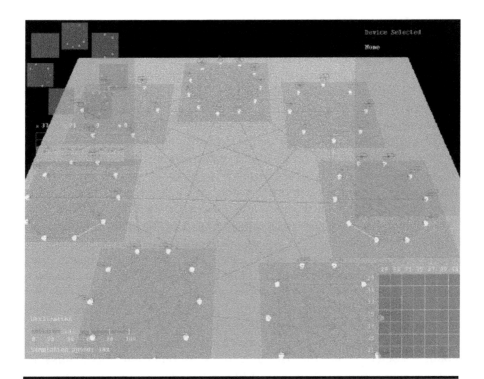

Figure 7.3   Logical view of residence network in Visual Network, showing traffic levels between and within different VLANs. (From Burns et al. (2003) Computer Networks, 43: 369-388. ©2003 Elsevier.)

Abstract Function Display

The Abstract Function level (shown in Figure 7.3) shows link utilization values. Link utilizations are shown for both physical and logical links, to balance the information across LANs and VLANs. Figure 7.3 shows only the logical links between VLANs, though the display can be configured to show physical links.

────────────────── ▼▲▼ ──────────────────

EYE ON VISUALIZATION

This display (Figure 7.3) uses:

1. Two levels of network structure
2. A matrix structure to show links in the lower right corner
3. Three dimensions
4. Polygon graphics at each network node
5. Polygon graphics used as mass data display symbols
6. A semi-transparent overview display in the top left corner

────────────────── ▲▼▲ ──────────────────

When the traffic load is light and utilization levels are low, the links appear with a dull green color. However, when the levels cross a warning threshold, which signifies that the amount of traffic is starting to significantly affect the data transfer rate through that link, the link will turn yellow. Links turn a salient red when they are nearly saturated with high amounts of traffic that result in severe network congestion and delays in transmission.

To aid in determining the traffic load that flows across the network between different LANs or VLANs, a utilization matrix is shown in the lower right-hand corner of the display. This display allows the network manager to quickly get a feel for where the traffic load is heaviest and what the distribution of the load is like between each pair of VLANs. In the combined mode, the display appears as a matrix with only its lower diagonal area filled in by a set of colors that match the colors of the corresponding links in the main view. Each cell of the matrix thus represents the combined level of traffic that is flowing in both directions between the two VLANs, represented by the identifier in both the row and column of the matrix. In Figure 7.3, for example, there is a single link that shows a critical amount of traffic flowing between VLAN 33 and 39, which is represented in the matrix as a red square located at the column labeled 33 and the row labeled 39. All other inter-VLAN links are operating in the normal range, so the other cells in the lower diagonal of the matrix show a green color.

Generalized Function Display

For the prototype display implementation, six variables or metrics were selected to provide information at the Generalized Function level: broadcasts, multicasts, errors, packets, octets, and utilization. The first five metrics can be directly queried from the management information variables (MIB) variables that are collected from an RMON agent for each port or interface of every network device being monitored. The last metric, utilization, can be derived from the other MIB variables to provide an indication of the percentage of the total link capacity that is currently being taken up by traffic passing in and out of a specific port on a network device.

In the prototype EID display, these six metrics are arranged in a polygon that appears above every network device on the main 3D view. The diagram shows the value of each variable represented along one axis of the polygon, with the center of the star representing the zero point for all variables and the ends of each axis representing the normalized

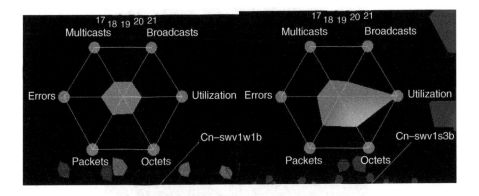

Figure 7.4 Two close-up shots of the polygon graphics. (From Burns et al. (2003) Computer Networks, 43: 369-388. ©2003 Elsevier.)

maximums for each metric. The current values for each metric are then plotted in appropriate positions along their respective axes, and the points are joined together to form an emergent shape or polygon, which can be quickly read or interpreted. This is illustrated in Figure 7.4, which shows two examples of a polygon graphic: one representing a normal situation (left) and the other (right) having at least one metric with a value that is in a critical state.

While every network device in the main view appears with a polar star diagram located immediately above it, only the selected object will show more detailed graphs stacked above the polar star. For example, a second polygon display was also used to show port information for the selected metric for each device, with trended information being displayed in a bar graph depicting current and past values for the selected port at the very top (see Figure 7.5).

Physical Function Display

The Physical Function display provides device-specific information that allows the network manager to get an idea of the capabilities of the equipment and components that physically comprise the network. In the prototype display, device-specific information is given in a semi-transparent panel located in the upper right-hand corner of the display and overlaid on top of the main view, similar to the overview map. This display provides some general information on the currently selected device, such as its name, IP address, device type, and number of ports, as shown in Figure 7.6.

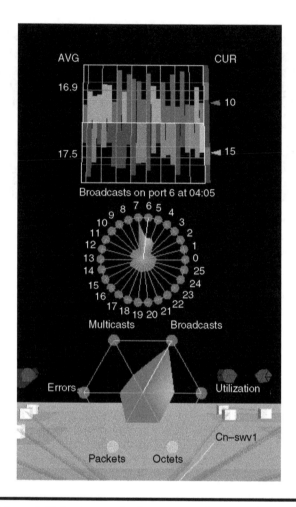

Figure 7.5 View of all the GF graphs that appear above a selected device. (From Burns et al. (2003) Computer Networks, 43: 369-388. ©2003 Elsevier.)

▼▲▼

EYE ON VISUALIZATION

This display (Figure 7.5) uses:

1. Two different polygon graphics, one with 6 variables, oen with 25 variables
2. Filled polygons that change color as they cross thresholds
3. Labels that are also symbols; they become brighter if there is a problem
4. A trending bar graph with colored bars that has some mass data features

▲▼▲

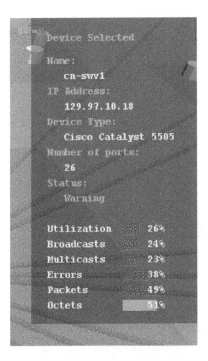

Figure 7.6   Close-up view of the device info panel. (From Burns et al. (2003) Computer Networks, 43: 369-388. ©2003 Elsevier.)

EYE ON VISUALIZATION

This display (Figure 7.6) uses:

1. Yellow, to make problems more salient
2. Semi-transparency; the network is still visible behind the display
3. Meters with digital readouts

In addition to information related to the properties or characteristics of the selected device, the device info panel also provides extra information concerning the device's current status. Repeating the information from the Generalized Function level enforces the means-end link between these levels.

Physical Form Display

A critical information requirement for network managers trying to solve a problem is the physical location of equipment, wires, and components

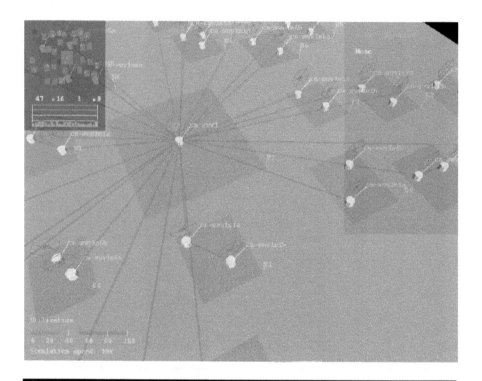

Figure 7.7 Screenshot showing physical topology and layout of devices within the Village 1 Residence network. (From Burns et al. (2003) Computer Networks, 43: 369-388. ©2003 Elsevier.)

that comprise the network. Without information concerning, for example, the location of failed equipment or the exact location of a port on a switch that is causing problems, a network manager would be helpless in trying to resolve any network problems that come up.

In the prototype display, topological information containing physical locations of devices and how they are interconnected is integrated into the main 3D view of the display. Network devices that appear on the map are arranged in approximately the same geographical locations as they would be in the real network. To portray this information more clearly, the network objects are placed within a number of different areas, shown as blue rectangles on the 3D plane in the main view, with each area representing a separate building in which the devices are physically located. This is illustrated in Figure 7.7, which shows the topology of the Village Residence network.

The blue areas are arranged on the map to reflect the geographical layout of the buildings in the real world, so that it is easy to determine where the devices are located based on which area they are in. The areas are also labeled with the names of the buildings, to allow the network manager to identify the areas in case it is not clear which buildings they represent based on their relative positions on the map.

In the main 3D view, links are also drawn between each device that is physically connected to other devices, providing important connection information at the physical level. The network manager can thus use this information to determine how the devices are physically connected to each other, which may be very important information when trying to trace a problem, such as a degradation in performance in a particular part of the network caused by a faulty wire.

Evaluation

This network management tool was compared against an industry quality tool, HPOpenView Network Node Manager (NNM). HPOpenView is a tool for broadband network management that is used by many organizations with large networks to monitor.

The two displays were compared in two tasks: the detection of network problems and the diagnosis of these problems. In both cases, the tools displayed data captured from our university network and saved for display in the tools. The two tools will be referred to as NNM and EID in the following discussion. On the detection tasks, faster detection times were measured with NNM than with the EID display. On all three diagnosis tasks, though, the EID display generated faster diagnosis times and more accurate diagnoses than NNM. A limited field study was also conducted with network managers at the university. Responses were positive, particularly towards the polar star graphic that was perceived to be very useful for problem diagnosis. There were concerns over how well this display would scale to a larger network situation.

AN ALTERNATIVE DESIGN: DESIGN FOR SCALABILITY

A project directed at scalability and network management occurred at the University of Toronto (Duez and Vicente submitted). This project had several goals: first, to apply WDA to a different kind of network, in this case a full campus network; second, to investigate how the resulting display supported operator monitoring under different levels of fault frequency; and third, to investigate visual ways of handling networks of large scale.

Work Domain Analysis

The work domain model in this project had many similarities to the previous work domain model (see Table 7.1). In Table 7.2, we show the

Table 7.2   Work Domain Model and Information Requirements for Network Management, Adapted from Duez and Vicente (submitted) and Duez (2003)

Level	Analysis	Variables
Functional purpose	Fast, reliable communication	% of active uncongested nodes
Abstract function	Conservation of information	Qualitative description of traffic load (normal, high, critical)
Generalized function	Processes	Load per port, ratio of good to bad packets, packet sizes, destinations, and errors over time
Physical function	Equipment capabilities	Device configuration and capacity, IP address
Physical form	Physical attributes of equipment	Model of the device, physical location, and contact person

general content of the model, with the variables used in the design in the final column.

If you compare this analysis with the previous analysis in Figure 7.1 and Table 7.1, you should notice some similarities. Both analyses focus on communication speed and quality at the Functional Purpose level. In both cases, information flow is the fundamental principle shown at the Abstract Function level. Network process statistics appear at Generalized Function, and capabilities and physical attributes at Physical Function and Physical Form.

Display Design

Figure 7.8 shows the resulting display. The main display uses a hyperbolic tree to show the main structure of the network. The hyperbolic tree is well designed for displaying systems with many nodes, as it selectively magnifies the area of interest without completely hiding the other nodes.

The network health view, in the top left corner, is designed from the Functional Purpose information. It uses the hyperbolic tree again, but in this case with small dots to represent the devices. In this way, the graphic is actually a Mass Data Display organized on a hyperbolic tree structure. This view would compare with the overview display in Figure 7.2. While both use the colored dot to obtain a mass data visualization, the hyperbolic tree provides additional information management that should help with larger scale networks. Physical form information is shown in the window directly

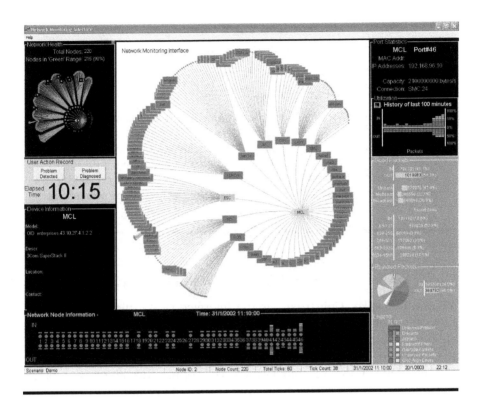

Figure 7.8   An alternative EID display for network management.

below. The windows along the right-hand side of the display show packets, errors, and utilization. Instead of the polar star graphic, bar graphs and pie charts are used to convey the same information. A time history is also shown.

―――――――――――――――――▼▲▼―――――――――――――

EYE ON VISUALIZATION

This display (Figure 7.8) uses:

1. A hyperbolic tree to show the structure of the network
2. A hyperbolic tree to organize a mass data display (top left corner)
3. A mass data display that serves as an overview (top left corner)
4. A pie graph to show packet quality (lower right corner)

―――――――――――――――――▲▼▲―――――――――――――

THE ANALYSIS IN THE DESIGN

Analysis: Table 7.2

This display shows four WDA levels:

FP: Fast reliable communication Hyperbolic tree overview, top left corner

AF: Conservation of information Good packet/rejected packet comparison, right corner

GF: Processes networking Packet pie chart, utilization history

PFn: Device information Device details, node info lower left and lower

ANALYSIS FOR RADIO COMMUNICATIONS

A Control Display Unit (CDU) is the interface to the flight management subsystems of aircraft and assists in navigational, mission planning, and radio communication activities. This project by Chéry et al. (1999) looked at a redesign of a CDU interface for a CH-146 Griffon Helicopter. The current CDU of the CH-146 Griffon Helicopter was comprised of an electroluminescent display screen with adjacent keys and an alphanumeric keyboard. Issues of concern with this interface included pilot workload and the ability to communicate and navigate effectively with the unit. This particular project examined communication with the CDU.

Work Domain Analysis

Chéry et al. (1999) developed an Abstraction Hierarchy for the radio communication work domain. This represented a subset of the functions of the CDU and did not include the navigation functions. They used five constraint levels and one level of decomposition. Table 7.3 shows their basic work domain model.

Display Design

Figure 7.9 shows an ecological display that was designed for the unit. The display has several features. At the top of the display, there is a display of how far the communication system can reach. This implements the Functional Purpose of the work domain model by letting the operator know which parties they can reach. The Venn diagram to the right of this display shows the amount of information transmitted, which implements

Table 7.3 Work Domain Model for Radio Communication

Level	Description
Functional purpose	Permit voice communication between remotely located parties
Abstract function	Communication theory, entropy of source, channel, equivocation, encoder, decoder, and destination
Generalized function	Signal transduction, frequency generation, interference generation, modulation, radiation, propagation, extraction, refraction, demodulation, absorption, attenuation, interference reduction, frequency conversion
Physical function	Microphone, electric source, ionosphere, terrain, airspace, antenna, filter, modulator, demodulator, oscillator, correcting network, speaker
Physical form	Appearance and physical connections

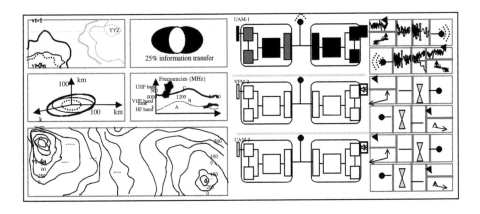

Figure 7.9 Ecological display for the radio control unit. (From Chéry, S., Vicente, K. J., and Farrell, P. (1999) Perceptual control theory and ecological interface design: Lessons learned from the CDU, *Proceedings of the 43rd Annual Meeting of the Human Factors and Ergonomics Society,* **389–393.)**

the Abstract Function level. Below these displays are two process displays, one for attenuation or information losses in different directions, and the other for information absorption by different levels of the atmosphere. To the right of the display is information on the activity on three channels, displaying the Physical Function information.

EYE ON VISUALIZATION

This display (Figure 7.9) uses:

1. A Venn diagram to show information transfer (top left)
2. A map to show radio coverage as an overview (top left corner)
3. Signal diagrams to show working channels (right side)
4. Mimic diagrams of components (center)

THE ANALYSIS IN THE DESIGN

Analysis: Table 7.3

This display shows four WDA levels:

FP:	Permit voice communication	Location overview of communication range top left corner
AF:	Principles of radio communication	Information transfer Venn diagram
GF:	Signal processes	3D range graphic power vs. frequency graph, topographical map
PFn:	Component connections	Symbolic channel maps (center), signal transfer graphics right side

Evaluation

Chéry et al. evaluated their display in comparison with an alternative display that was under consideration at the time. They performed a cognitive walkthrough of the two displays, examining normal and abnormal operations with the display using a realistic mission-based scenario. Although not a formal evaluation, the comparison revealed some of the possible strengths of the EID display. This brief analysis suggested that the EID display should be able to support performance in both normal and abnormal situations. The display has yet to be formally tested.

HANDLING THE CHALLENGES

Challenge 1: Determining System Boundary

The network management examples have inherent system boundary issues due to the connectivity of the situation. In both examples by Kuo and

Burns (2000), it was decided to limit the network under analysis. They both did this by ending the network at a connection to a main switch. This created a physical system boundary.

The second way of managing the system boundary was to transform the Part-Whole Decomposition from device aggregation to information aggregation. The information dimension is more general and will hold across all sizes of networks. This transformation makes the model applicable to all networks following OSI layers. The alternative would have been to analyze "network" then decide subnetworks, then possibly workgroups and devices. In particular, because this was an Ethernet LAN environment with both physical and virtual subnetworks, this physical bounding did not suit the problem.

Challenge 2: Determining WDA Content

When you apply WDA to systems that are not based on mass and energy relations, you need to determine other principles behind the system. This is why, in these examples, information flow is the main flow modeled at the Abstract Function level. The network management example models data transfers between network layers, and the radio communication example models information transfer between a sender and receiver. While physically there is a medium for signal transfer in both cases, this has not been included in the analyses.

You will see many similarities between the two models when modeling processes. Generation, transmission, and reception occur in both systems, particularly at the signal level. The network management example also shows processes at higher levels of information aggregation.

As in the other work domain models we have investigated, these models describe the devices. From Physical Function to Physical Form, the models are similar to other work domain models that we have seen.

Challenge 3: Developing a Diagnostic Graphic for a Low-Capacity Situation

In many production-based systems, the objective is to move material through the system as quickly as possible, or as close as possible to optimum rates. A network operates quite differently. The primary measures of network health are traffic and capacity. In particular, lower traffic levels mean a healthier network. Rising traffic levels indicate a problem is developing. Operating with lots of spare capacity maximizes availability, minimizes errors, and minimizes delays, three purposes of our network from our work domain model. These principles hold during every process and when using every device in the network.

In developing a device graphic we wanted to capture this aspect of low traffic across multiple parameters. We used the polar star graphic because it provided integration of multiple parameters. In this case, though, we did not normalize the values, since the values should always be minimized. This means our polar graphic cannot be used in the configural way that a polar star is used. However, it has an emergent feature of a point when any values get high, which makes the graphic diagnostic.

Challenge 4: Using a 3D Ecological Display

Ecological displays are just a method for designing displays and can use many different visualization approaches. In this project we developed a three-dimensional display in order to show the entire network and the many levels of information on a single display. We used the topology, from the Physical Form level, as the base of the display and put higher level information above the topology. The advantage is that this display format allows multiple levels of information to be displayed. The disadvantage is that information can be obscured, which is often a problem with 3D displays, and information management becomes a physical navigation task for the user. These trade-offs need to be considered when designing 3D ecological displays.

Challenge 5: Developing Visualizations That Scale with Large Amounts of Data

The display by Duez shows one way of developing an EID that scales with large amounts of data. The hyperbolic tree manages the display of the information, without ever totally obscuring the data. Duez used the tree again to create an overview display with mass data properties. Kuo used the same concept of adapting the main display into an overview display. It probably will not scale as effectively as Duez's display, but demonstrates the same concept. In both cases, the main display was reduced in size and the primary objects converted to a symbol (a circle in both cases). The symbol undergoes a simple color change with the state of the object, calculated off some threshold of the other variables. The large number of symbols, with the simple color change, creates a Mass Data Display that can handle more information, although only at an overview level, than the main graphic. Both of these concepts could be developed further, possibly using these overview graphics as navigation tools.

8

MEDICAL SYSTEMS

Medical systems help caregivers monitor and treat patients at various levels of health and disease progression. Medical practitioners including doctors, nurses, nurse practitioners, interns, and other caregivers use these systems in various settings to plan and coordinate prescribed therapies that will minimize risks to the patient while maximizing the intended outcome.

The demand for providing quality care for sicker patients has increased over time. In response, patient monitoring and therapeutic equipment have increased in number and complexity. Hospitals are full of advanced and configurable technologies that are utilized in the operating room, intensive care units, and therapy wards.

At the same time, there is a trend for health care to become more collaborative and distributed, especially for less-critical conditions. Instead of patients coming to caregivers in central locations, there is an increasing demand for telehealth, remote care, or self-diagnosis as an option. This trend has emerged from the advantages of keeping people at home and the economic benefits of managing health care resources more effectively. Remote monitoring and therapeutic technologies are seen as enabling this societal need.

As a result of the above trends, the demands on medical personnel, both at the hospital and in managed distributed care, will continue to increase because of the associated additional monitoring and supervisory tasks in this highly dynamic environment. Despite numerous technological advances, concerns and limitations have been documented. A report from the Institute of Medicine (2000), *"To err is human,"* has noted that a key contributor to human errors in health care is poor device design that makes it easier for people to get confused and make mistakes. One of the key recommendations is the use of human factors in design to provide feedback to the user on medical device constraints. In addition, having more feedback on patient condition could lead to better patient care and fewer errors. Both will need to be realized to help meet the emerging health care needs.

CHALLENGES

EID provides a unique perspective to help define what the information requirements are to better support individual patient care and collaborative, coordinated medical care. There are several distinct challenges, though, in applying EID in this domain:

- How to choose the system boundary
- How to decide when to stop decomposing of the Part-Whole Hierarchy
- How to integrate medical information
- How to deal with issues associated with sensor limitations and availability
- How to determine information requirements for different roles in the medical environment
- How to extend EID to other sensory modes (e.g., auditory)

Three case studies are presented that look at different medical environments to help address these challenges: newborn monitoring in the neonatal intensive care, patient monitoring in the operating room, and diabetes management.

CHALLENGES WITH MEDICAL SYSTEMS

Deciding the system boundary
Deciding the end point of the Part-Whole Analysis
Integrating medical information
Working with low levels of sensor availability
Allocating team members to different parts of the work domain
Extending EID to other sensory modes

DESIGN FOR OXYGENATION MONITORING IN THE NEONATAL INTENSIVE CARE UNIT

Sharp and Helmicki (1998) conducted a WDA of neonatal intensive care and developed displays based on EID to assist care providers in assessing neonatal oxygenation. This case study also presents a summary of their findings as compared to an existing display.

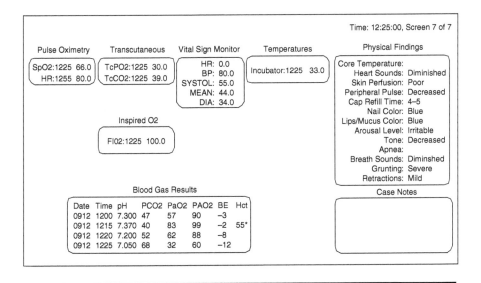

Figure 8.1 An existing display for the neonatal intensive care unit. (Reprinted with permission from Sharp, T. D. and Helmicki, A. J. (1998) The application of the Ecological Interface Design approach to neonatal intensive care medicine," *Proceedings of the 42nd Annual Meeting of the Human Factors and Ergonomics Society, 350–354.* ©1998 by the Human Factors and Ergonomics Society. All rights reserved.)

The neonatal intensive care unit has numerous monitors that are driven by individual sensors. These monitors are used to assess patient condition and reactions to interventions. Within the unit, a central station typically monitors a number of newborns who are in critical condition. Medical practitioners monitor the newborns frequently and make therapeutic adjustments or other interventions in response to or in anticipation of treatment needs. These practitioners need to detect deviations from stable conditions as early as possible, to ensure there is enough time to correctly identify, diagnose, and manage problems before a newborn's life is in jeopardy. Figure 8.1 shows an existing display for neonatal oxygenation (Sharp and Helmicki 1998), consisting of individual instrument readings, a table of recent blood gases, physical findings, and case notes.

System Boundary

A critical function important to life in the neonatal intensive care unit is oxygenation. Sharp and Helmicki (1998) defined the body systems and functions associated with neonatal oxygenation as the work domain for the purposes of designing new displays for the neonatal intensive care

Decomposition			
Whole ───────────────────────────────► Parts			

		Body	Body Systems	Body Organs	Cellular
Abstract	Purpose	Homeostasis: Maintain internal environment			
Abstraction	Balances	Balance oxygen supply and demand	Oxygen balances in respiratory, cardiovascular and metabolic systems	Oxygen balance in alveoli, pulmonary blood, arterial blood, capillary blood, and tissues	Balance oxygen supply and demand at the mitochondria of all cells in the body to support anaerobic metabolism
	Processes	Homeostatic processes	Oxygenation, ventilation	Alveolar ventilation	
	Transport, Storage, and Control			Homeostatic response, vascular volume, vasoconstriction, blood pressure, blood flow	
Concrete	Physical Form			X-rays: lungs, heart, size and location	

Figure 8.2 A work domain model for tissue oxygenation. (From Sharp, T. D. (1996) Ecological interface design for the neonatal intensive care unit, unpublished thesis, University of Cincinnati. Reprinted with permission.)

unit. Sharp and Helmicki created a general work domain map using the Abstraction Hierarchy (AH) for the body systems involved with oxygenation.

Work Domain Analysis

A WDA was performed for the neonatal intensive care unit (Figure 8.2). The Part-Whole Hierarchy is organized at body, body systems, body organs, and cellular levels. Below is a description of the various levels of the AH.

Purpose: The top level of the AH refers to the maintenance of homeostasis or the internal environment of newborns. Homeostasis refers to keeping a biological system in balance and in condition to maintain life. Disturbances in body functions as described in the lower levels of the AH may cause deviations from homeostasis.

Balances: To maintain the internal environment, the demand for and supply of nutrients must be balanced. This level models the balance of these nutrients, in particular the oxygen/carbon dioxide balances.

Processes: This level includes the processes needed to support the balances required for adequate oxygenation and ventilation. These processes correspond to regulated flows of oxygen and removal of carbon dioxide.

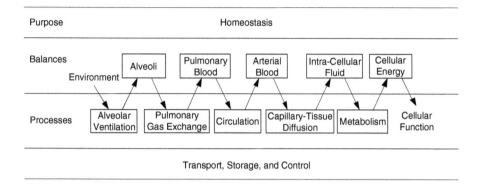

Figure 8.3 The higher levels of the work domain model. (From Sharp, T. D. (1996) Ecological interface design for the neonatal intensive care unit, unpublished thesis, University of Cincinnati. Reprinted with permission.)

Transport, Storage, and Control: This level includes components that support the processes. They include transport functions (e.g., network of blood vessels), storage capabilities (e.g., storage of oxygen in hemoglobin), and control (e.g., the ability to generate adequate blood flow).

Physical Form: This level includes the actual arrangements and interconnections of the various body subsystems.

The portion of the AH shown in Figure 8.3 depicts the means-ends links. This representation was used to create more detailed models of processes and balances that drove the design of the EID display.

From the representation in Figure 8.3, Sharp (1996) created a number of causal models at various levels of the Abstraction Hierarchy. Five models are presented here for illustration. Figure 8.4 represents the alveolar ventilation process. Figure 8.5 represents the pulmonary gas exchange process. Figure 8.6 represents the circulation process. Figure 8.7 represents the capillary diffusion process. Finally, Figure 8.8 represents the processes associated with metabolism. Key relationships were derived with these models and were used in creating the EID display (Figure 8.9).

Instrumentation Availability

Two challenges with medical domains are that they are largely missing sensors (because they are too invasive or not available) and physiological knowledge is continuously changing and improving. In many cases, most of the information identified by a work domain model will not be available to practitioners with these medical systems. This makes the strategic selection of sensors and instrumentation availability analysis important in these systems.

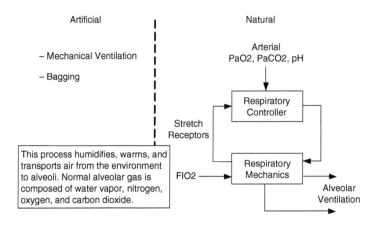

Figure 8.4   The model of the alveolar ventilation process. (From Sharp, T. D. (1996) Ecological interface design for the neonatal intensive care unit, unpublished thesis, University of Cincinnati. Reprinted with permission.)

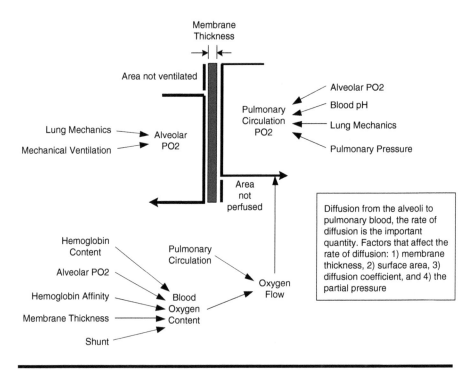

Figure 8.5   The model of pulmonary gas exchange. (From Sharp, T. D. (1996) Ecological interface design for the neonatal intensive care unit, unpublished thesis, University of Cincinnati. Reprinted with permission.)

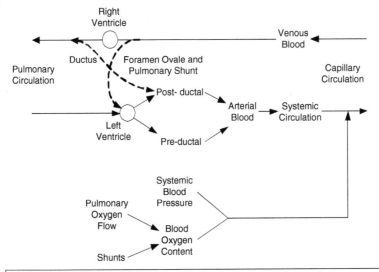

The role of the circulatory system is to transport oxygen (and other nutrients) from the pulmonary systems to the tissues that need it. There are two factors that influence this: 1) the ability of the system to provide a consistent blood pressure, and 2) the existence of shunts in circulation.

Figure 8.6   The model of the circulatory system. (From Sharp, T. D. (1996) Ecological interface design for the neonatal intensive care unit, unpublished thesis, University of Cincinnati. Reprinted with permission.)

As in the typical information availability analysis, Sharp and Helmicki identified directly sensed variables, analytically derived variables, and variables that could not be obtained. They also used a fourth category they called "heuristically mapped" variables. These were variables that were qualitative and subjective and were determined by discussions with medical practitioners. Table 8.1 (derived from Sharp 1996) shows examples of the variables associated with the four categories.

Display Design

The display for the neonatal intensive care unit, shown in Figure 8.9, was developed by mapping the information requirements generated by the work domain model onto appropriate display forms. In creating the display mapping, forms were selected that would allow the practitioners to use their perceptual abilities as much as possible. The mapped information requirements were then organized on the display based primarily on the organization of the balances and process models (see Figure 8.4 through

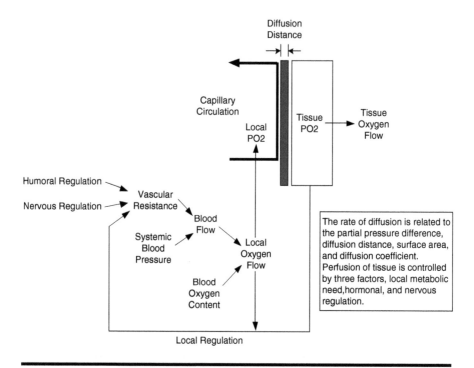

Figure 8.7   The model of capillary-tissue diffusion. (From Sharp, T. D. (1996) Ecological interface design for the neonatal intensive care unit, unpublished thesis, University of Cincinnati. Reprinted with permission.)

Figure 8.8). This can be seen in Figure 8.9; the top row of readings roughly corresponds to the balances level, the second row to the processes. Key variables and relationships that were measured, heuristically mapped, or analytically derived from the process models in Figure 8.4 through Figure 8.8 are also presented in the EID display.

Sharp (1996) identified two limitations. The first limitation of the display was that some of the identified variables from the work domain model were not included on the display because no robust mapping could be found. In addition, there was concern about the robustness of variables that were mapped using analytical and heuristic methods. Medical practitioners may not understand the limitations associated with these mappings and may use the information presented when it is not appropriate. The second limitation relates to the difficulty in setting target or normal ranges for the identified variables. There is a high variability of infant conditions, and generic ranges cannot be set for all cases. These ranges are typically set by care providers after they have made their assessments on patient state. Target values for physiological variables are not fixed constants and

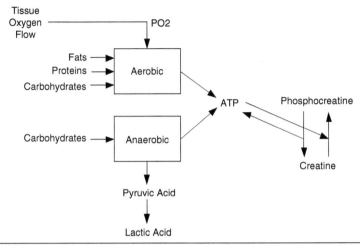

When the cellular PO2 is greater than 1 mmHg, the availability of oxygen is no longer a limiting factor in the rates of chemical reactions. In this case the cells can use aerobic metabolism to create ATP, the energy currency of the cell. When the oxygen level drops below this level aerobic metabolism is limited, causing anaerobic metabolism to be used to create the required ATP. The rate of these metabolic processes is controlled by the concentration of ATP.

Figure 8.8   The model of metabolism. (From Sharp, T. D. (1996) Ecological interface design for the neonatal intensive care unit, unpublished thesis, University of Cincinnati. Reprinted with permission.)

may not even have a generic value; many factors influence their ranges. Medical practitioners decide what the target range is for each infant.

Evaluation

Sharp and Helmicki (1998) tested the differences between the current display used at a neonatal intensive care unit (shown in Figure 8.1) and the EID display (shown in Figure 8.9). Sixteen physician participants used both displays in a within-subjects design, alternating the order of use of the displays. The participants had three levels of expertise: residents, fellows, and attending physicians. The participants were expected to diagnose four simulated scenarios showing common acute situations.

Physicians performed better with the EID display on a task where they had to select the best diagnosis from a set of possible diagnoses. There was a significant interaction between the displays and the level of experience, with the EID display improving performance most with the least-experienced group (the residents). Only 1 subject out of the 16 performed better with the existing display, despite it being the more familiar display.

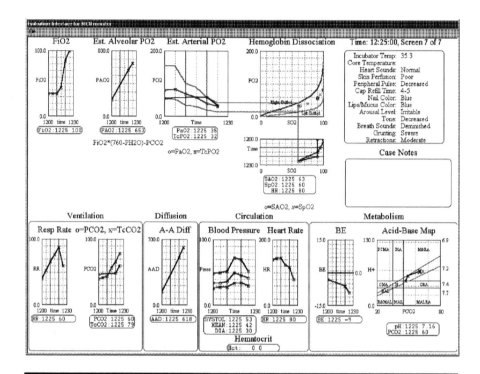

Figure 8.9 An EID display for the neonatal intensive care unit. (Reprinted with permission from Sharp, T. D. and Helmicki, A. J. (1998) The application of the Ecological Interface Design approach to neonatal intensive care medicine," *Proceedings of the 42nd Annual Meeting of the Human Factors and Ergonomics Society,* 350–354. ©1998 by the Human Factors and Ergonomics Society. All rights reserved.)

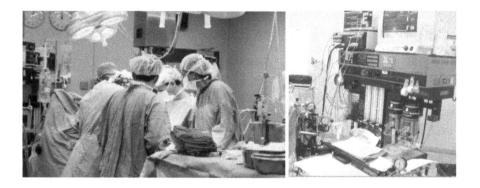

Figure 8.10 The operating room environment: surgical team (left) and anesthesia workstation and monitors (right). (Courtesy of Dr. John Doyle, the Toronto Hospital.)

Table 8.1   Information Availability Analysis for Tissue Oxygenation

Category	Variable in the AH	Actual Measurement
Directly sensed variables	Environmental oxygen concentration (balance level)	Inspired oxygen concentration, FiO_2, directly measured
Analytically derived variables	Balance in the alveolar PO_2 (balance level)	Analytical model exists for estimating its value; physicians calculate by hand or with computer
Variables that could not be obtained	ATP level in each cell in the body	Cannot be measured in clinical setting
Heuristically mapped variables	Ventilation process (process level)	Use arterial PCO_2 as a useful estimate of alveolar PCO_2 and therefore ventilation; CO_2 diffuses through the lung wall rapidly

EYE ON VISUALIZATION

This display (Figure 8.9) uses:

1. Graphs with thresholds and digital values
2. Variables plotted against a background of thresholds (hemoglobin dissociation)

THE ANALYSIS IN THE DESIGN

Analysis: Figure 8.2 through Figure 8.8

This display is only one of several and shows two WDA levels:

AF: Balances   Variables roughly in top row; connector lines between arterial PO_2, hemoglobin dissociation, and SO_2

GF: Processes   Variables roughly in bottom row; graphics plotted over time, acid-base map

PATIENT MONITORING IN THE OPERATING ROOM

One critical and dynamic work environment in medicine is the operating room. Teams of medical personnel work to stabilize patient conditions and perform medical procedures and interventions. This environment is highly variable, dynamic, and uncertain and densely populated with various technological tools (Figure 8.10). The medical team must adapt to changing resources and patient conditions. Similar to the previous case, current interface technologies typically display single-sensor information. Practitioners must integrate this information with experience and past patient history to determine the appropriate course of action. One of the challenges of this environment is that the work domain, the patient, is a "broken" system due to conditions before the surgery or during the actual surgery itself. The surgeon must maintain the broken system and work to fix it during the course of the surgery. The nature of the surgery and the health of the patient can greatly affect the complexity of this domain.

System Boundary

The operating room is a complex and interconnected environment involving the patient, the medical equipment, and the medical team of nurses, surgeon, and anesthesiologist. Figure 8.11 shows the various aspects of

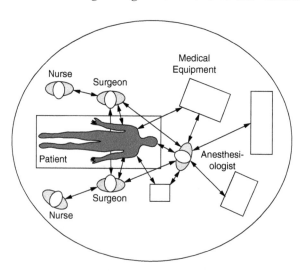

Figure 8.11 Operation room environment showing the system boundary. (Adapted from Hajdukiewicz, J. R., Vicente, K. J., Doyle, D. J., Milgram, P., and Burns, C. M. (2001) Modeling a medical environment: An ontology for integrated medical informatics design, *International Journal of Medical Informatics* 62: 79–99. ©2001, with permission from Elsevier.)

this environment. In this system, the system boundary could be defined in a number of different ways. The boundary could be limited to specific medical equipment, the patient, or the entire operating room. Each of these boundaries changes the scope of the problem and the nature of the solution. For a design focused on equipment diagnostics you would place your system boundary around the equipment. For a design geared towards team efficiency or planning, you might include the entire team in your boundary. In this particular case, the analysts chose to restrict their system boundary to the patient, shown by the rectangle in Figure 8.11.

Work Domain Analysis

The analysts here used the typical five levels of abstraction in the Abstraction Hierarchy and five levels of decomposition in the Part-Whole Hierarchy. The five levels of Part-Whole Hierarchy were Body, System, Organ, Tissue, and Cell. The work domain was not decomposed beyond the cellular level. A system was defined as a collection of organs (e.g., the cardiovascular system), an organ was defined as a collection of tissues, and a tissue was defined as a collection of cells.

The work domain model has the following abstraction levels:

Purposes: This level contained the physiological purposes governing the interaction between the patient and the medical environment. It included maintaining homeostasis (maintenance of internal environment), adequate perfusion, circulation, oxygenation, ventilation, and circulatory volume.

Balances: This level included physiological balances in salt and water, oxygen supply and demand, electrolytes, and general conservation relationships of mass, energy, and momentum.

Processes: This level contains generic physiological processes that are to be coordinated irrespective of the underlying physiology and component configuration. This level included circulation, perfusion, oxygenation, ventilation, metabolism, storage, diffusion, osmosis, binding, chemical release, and heat transfer.

Physiology: This level contains the organ- and component-related physiological functions available to establish and maintain the processes. Examples include the functioning of organs and other body components. This is also the level where actions can be performed to reconfigure physiological processes (e.g., administer a vasodilator to increase blood flow through the arterioles).

Anatomy: This level included specific anatomical structures. Examples include the location, appearance, form, and material structure of the human heart.

These levels were repeated at the five part-whole levels. Figure 8.12 shows the breadth of the analysis and the type of information at each part-whole level. The extension below the work domain model shows a partial work domain model for the cardiovascular system. Note the greater detail in the partial work domain model than in the overall view of the work domain model. The lower levels include the cardiac and circulatory functions necessary to support the higher level purposes of adequate circulation and blood volume; the higher levels provide reasons for lower level functions. Figure 8.13 shows the causal models for the cardiovascular system at the levels of balances and processes. This part of the analysis shows the flow patterns through the various systems and organs. Physiology level information is inherent in the labeling of the systems and organs in the figure.

▼▲▼

Analysis for a different system boundary.
The WDA below was performed for a ventilation system in the operating room by Sowb et al. (1998). The ventilation system is a set of equipment that is used to maintain oxygen levels in the patient during surgery. The analysis is more tightly restricted and gives details very specific to this problem. The patient is not in the boundary and not in the analysis, only the ventilation equipment. A designer using this analysis would focus their display on O_2 and N_2O pressures and flows and equipment diagnostics. Note also the return to traditional labels in the model, which matches well with this domain of equipment monitoring.

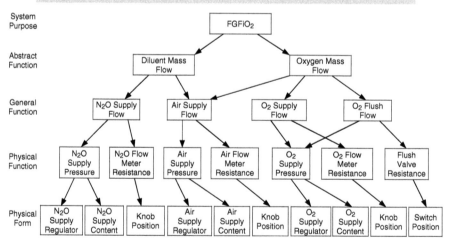

From Sowb, Y. A., Loeb, R. G., and Roth, E. M. (1998) Cognitive modeling of intraoperative critical events, *Proceedings of the IEEE Meeting on Systems, Man, and Cybernetics*, 2532–2538.

▲▼▲

Level of Aggregation

a)

		Body	System	Organ	Tissue	Cell
Level of Abstraction	Purposes	Homeostasis (Maintenance of Internal Environment)	Adequate Circulation, Blood Volume, Oxygenation, Ventilation	Adequate Organ Perfusion, Blood Flow	Adequate Tissue Oxygenation and Perfusion	Adequate Cellular Oxygenation and Perfusion
	Balances	Balances: Mass and Energy Inflow, Storage, and Outflow ★	System Balances: Mass and Energy Inflow, Storage, Outflow, and Transfer ★	Organ Balances: Mass and Energy Inflow, Storage, Outflow, and Transfer ★	Tissue Balances: Mass and Energy Inflow, Storage, Outflow, and Transfer ★	Cellular Balances: Mass and Energy Outflow, and Transfer ★
	Processes	Total Volume of Body Fluid, Temperature, Supply: O_2 Fluids, Nutrients, Sink: CO_2 Fluids, Wastes	Circulation, Oxygenation, Ventilation, Circulating Volume	Perfusion Pressure, Organ Blood Flow, Vascular Resistance	Tissue Oxygenation, Respiration, Metabolism	Cell Metabolism, Chemical Reaction, Binding, Inflow, Outflow
	Physiology		System Function	Organ Function	Tissue Function	Cellular Function
	Anatomy			Organ Anatomy	Tissue Anatomy	Cellular Anatomy

* Balances include: Water, Salt, Electrolytes, pH, O_2, CO_2

b)

		System	Subsystem	Organ	Component
Level of Abstraction	Purposes	Adequate Circulation and Blood Volume			
	Balances	Cardiovascular System: Mass Inflow, Storage, and Outflow	Pulmonary and Systemic Systems: Balances Mass Flows; Inflow, Storage, Outflow, and Transfer	Organ Vascular Network: Balances Mass Flows; Mass Inflow, Storage, Outflow, and Transfer	Vascular Components: Balances Mass Flows; Mass Inflow, Storage, Outflow, and Transfer
	Processes	Circulation, Volume, Fluid Supply and Sink	Pulmonary and Systemic Circulation (Pressure, Flow, Resistance) and Volume, Fluid Supply and Sink	Cardiac Output, Organ Circulation (Pressure, Flow, Resistance), Fluid Supply and Sink from each Vascular Network	Circulation through Vascular Components (Pressure, Flow, Resistance), Vascular Blood Volume, Fluid Supply and Sink
	Physiology	Cardiovascular System Function	Pulmonary and Systemic System Function	Cardiac Function (Heart Rate, Rhythm)	Atrial and Ventricular Function; Arterial, Arteriolar, Capillary, Venule, Venous Function
	Anatomy			Cardiac Anatomy	Atrial and Ventricular and Vascular Anatomy

Figure 8.12 Work domain model of a patient with expansion of the cardiovascular system. (Reprinted with permission of Hajdukiewicz, J., Doyle, D. J., Milgram, P., Vicente, K. J., and Burns, C. M. (1998) A work domain analysis of patient monitoring in the operating room, *Proceedings of the 44th Annual Meeting of the Human Factors and Ergonomics Society*, 1034–1042. ©1998 by the Human Factors and Ergonomic Society. All rights reserved.)

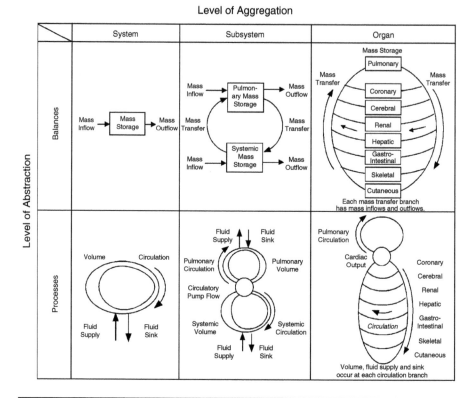

Figure 8.13 Causal map of part of the work domain model. (Reprinted with permission from Hajdukiewicz, J. R., Vicente, K. J., Doyle, D. J., Milgram, P., and Burns, C. M. (2001) Modeling a medical environment: An ontology for integrated medical informatics design, *International Journal of Medical Informatics* **62: 79–99. ©2001, with permission from Elsevier.)**

Information Availability

For this work domain, the analysts examined the information that was available, mapping it on to the work domain model (Figure 8.14). In this case, only a few variables could be sensed (e.g., electrocardiograph (ECG) waveform, heart rate, blood pressure, blood oxygen saturation, patient pulse, and skin color). Some indirect measures could also be derived (e.g., heart rate could be indirectly derived from the arterial blood pressure waveform). In some cases, the mappings between variables and sensors was not one to one, with some sensors providing more than one variable, and some variables being derived from several sensors.

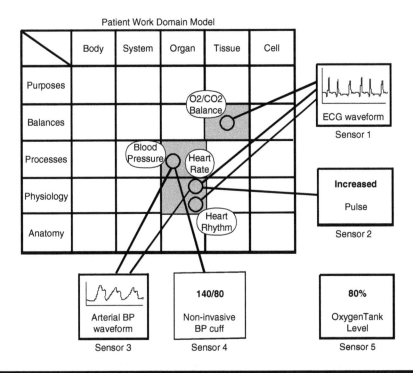

Patient Work Domain Model

Figure 8.14 Information availability analysis, operating room example. (Adapted from Hajdukiewicz, J. R., Vicente, K. J., Doyle, D. J., Milgram, P., and Burns, C. M. (2001) Modeling a medical environment: An ontology for integrated medical informatics design, *International Journal of Medical Informatics* 62: 79–99. ©2001, with permission from Elsevier.)

Function Allocation Using WDA

The work domain model can be used to understand the roles of different people working on the system. In this case, the model was used to identify two regions — that of the surgeon and that of the anesthesiologist. The surgeon is concerned with the workings and functions of various organs and tissues, whereas the anesthesiologist works at maintaining the conditions for homeostasis and eventual restoration of the patient to consciousness. As you can see in Figure 8.15, there is some overlap between their roles. This type of work domain-based functional allocation could be used to design different ecological displays for the surgeon and anesthesiologist, and define what information needs to be shared to improve their coordination and collaboration.

Level of Aggregation

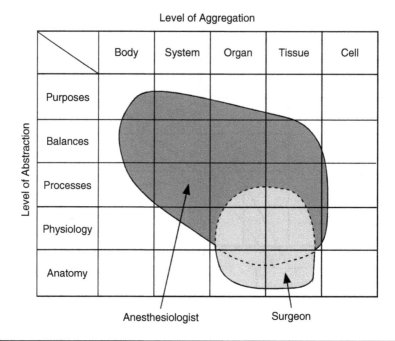

Figure 8.15 Functional allocation in the operating room using WDA. (Reprinted with permission from Hajdukiewicz, J. R., Vicente, K. J., Doyle, D. J., Milgram, P., and Burns, C. M. (2001) Modeling a medical environment: An ontology for integrated medical informatics design, *International Journal of Medical Informatics* 62: 79–99. ©2001, with permission from Elsevier.)

Display Designs and Evaluations

Three examples of displays are presented here. Two did not develop directly from the analysis above, but we have included them as examples of designs that meet some of the requirements in the work domain model in Figure 8.12 and Figure 8.13. In Chapter 4, we discussed selecting existing designs that meet the requirements of a work domain model using the Functional Information Profile. The third example was motivated by EID and other approaches from Cognitive Work Analysis (Vicente 1999) and used sonification as the display.

Blike's Display

With the first display (Figure 8.16), Blike created a graphical object display based on a cognitive analysis, with the goal of enhancing the accuracy and efficiency of an anesthesiologist in diagnostic tasks (Blike et al. 1999; Zhang et al. 2002). The diagnostic tasks studied were recognition and

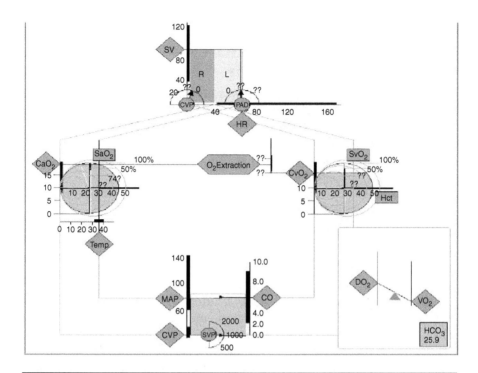

Figure 8.16 Blike's display for recognition of shock under anesthetic. (Reprinted with permission from Zhang, Y., Drews, F. A., Westenskow, D. R., Forest, S., Agutter, J., Bermudez, J. C., Blike, G., and Loeb, R. (2002) Effects of integrated graphical displays on situation awareness in anaesthesiology, *Cognition, Technology & Work,* 4:82–90. ©2002 by Springer-Verlag.)

differentiation of shock types: anaphylaxis, bradycardia, myocardial ischemia, hypovolemia, pulmonary embolus.

------------------------------▼▲▼------------------------------

Eye on Visualization

This display (Figure 8.16) uses:

1. Geometric features with scales showing the relations between key variables within a graphic element
2. Connector lines between graphic elements to show relations across variables
3. Connector lines to show the state of variable balances (bottom right side)

------------------------------▲▼▲------------------------------

▼▲▼

THE ANALYSIS IN THE DESIGN

Analysis: Figure 8.12 and Figure 8.13

Although this display (Figure 8.16) was not generated with the WDA in Figure 8.12 and Figure 8.13, the variables used in design and their relations can be mapped onto the WDA. This display is only one of several and shows three WDA levels:

AF: Balances   Balance between oxygen and supply and consumption (bottom right side)

GF: Processes   O$_2$ extraction (center), circulatory function (bottom center)

PF: Physiology   Heart function graphics (top)

▲▼▲

As shown in Figure 8.16, the object display provides information about stroke volume (SV), pulmonary artery pressure (PAD), heart rate (HR), arterial blood oxygen content (CaO$_2$), arterial blood oxygen saturation (SaO$_2$), temperature, oxygen extraction, mixed venous oxygen content (CvO$_2$), mixed venous oxygen saturation (SvO$_2$), hematocrit (Hct), mean arterial blood pressure (MAP), systemic vascular resistance (SVR), cardiac output (CO), oxygen delivery (DO$_2$), oxygen consumption (VO$_2$), and bicarbonate (HCO$_3$). The display contained more specific information about the cardiovascular system than typically used by anesthesiologists.

The above variables were derived or measured directly or indirectly. The variables were also integrated based on causal models of cardiovascular function. Five measured variables (HR, BP, CVP, PAD, and CO), two derived variables (SV and SVR), four relationships {(MAP − CVP) = CO * SVR; CO = HR * SV; LVEDV * PAD; RVEDV * CVP}, and the decision model that partially maps onto the work domain model were incorporated into the display shown in Figure 8.16.

Graphical features include labels pointing to the variables available, the current values, the "normal" data ranges (i.e., alarm limits) represented as black bars on the reference scales, and a dynamic lower alarm limit for MAP defined by the function {LIMIT = CVP + 55}. The interconnections and graphical features represent the relationships among the identified cardiovascular variables. For example, DO$_2$ and VO$_2$ have a line connected between the two; when oxygen delivery and consumption is balanced, the line should be approximately horizontal.

In one study by Blike et al. (1999) this display was designed to show functional cardiovascular physiology by integrating related hemodynamic variables. Evaluations of this display using static pictures found a performance improvement in the anesthesiologist's ability to recognize and rapidly respond to cardiac events. They also found an improvement in the accuracy of diagnoses. However, for this study, it was unknown whether the display also helps to increase the medical practitioner's overall situation awareness during the type of dynamic situation represented in real practice. An increase in situation awareness would suggest that the display might also improve the treatment of critical events.

Another study by Blike et al. (2000) was designed to measure situation awareness. The display was found to improve situation awareness for one of four test scenarios. Low-level situation awareness was higher with the traditional display (i.e., individual patient variables), and medium-level situation awareness was higher with Blike's display.

3D Integrated Display

The second display, shown in Figure 8.17, is an alternative to Blike's display that integrates medical functional information (Zhang et al. 2002). This display shows an integrated 3D representation of patient variables. The display is composed of four separate interactive windows, each one

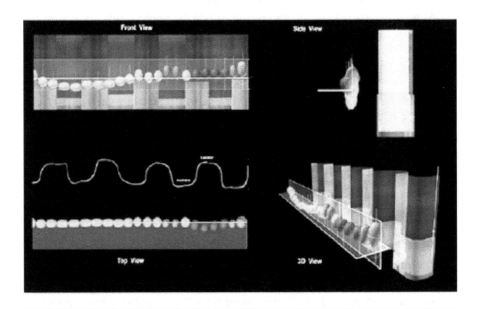

Figure 8.17 A 3D integrated display. (Reprinted with permission from Zhang, Y., Drews, F. A., Westenskow, D. R., Forest, S., Agutter, J., Bermudez, J. C., Blike, G., and Loeb, R. (2002) Effects of integrated graphical displays on situation awareness in anaesthesiology, *Cognition, Technology & Work,* 4:82–90. ©2002 by Springer-Verlag.)

showing different angles to the patient information. The 3D perspective view provides an overall integrated display of all recorded physiological data. Front, side, and top views show the same data from different angles to better see changes in cardiac state; these views provide more precise graphic forms of the patient data. For example, the side view provides trend information associated with the cardiac function; the front view provides information about the actual cardiac state.

The display shows many of the variables monitored during surgery. Eight variables are shown in real time on a single screen (heart rate; systolic, diastolic, and mean blood pressure; arterial oxygen saturation; respiratory rate; and inspired and expired gas content). Normal reference grids are provided, supporting the detection of changes in patient state; deviations from the normal condition are clearly visible from changes in the graphic forms in relation to the reference grids. For example, a change in cardiac output is presented by an increase of the height of the heart object (see Figure 8.18). The display also integrates traditionally isolated physiological data (e.g., cardiac output and heart period). The functional interactions among variables are displayed graphically; many of these relationships could be derived from a WDA.

Animation was used to help the medical practitioner interpret the physiological data, compare the monitored variables with the expected normal values, and come to the correct conclusions. The display also highlights physiological relationships between monitored variables through the object design, object location, and display structure. For example, the cardiac object (sphere) changes shape with each heart beat; its height is proportional to the cardiac volume, and its width is proportional to heart beat period. Even though these displays were not motivated directly by EID, the patient variable and functional relationships could be derived from the work domain model shown in the previous examples.

▼▲▼

THE ANALYSIS IN THE DESIGN

Analysis: Figure 8.12 and Figure 8.13

Although this display (Figure 8.17 and Figure 8.18) was not generated with the WDA in Figure 8.12 and Figure 8.13, the variables used in design and their relations can be mapped onto the WDA. This display is only one of several and shows two WDA levels:

GF: Processes   Cardiac output, blood pressure variables
PF: Physiology   Heart function

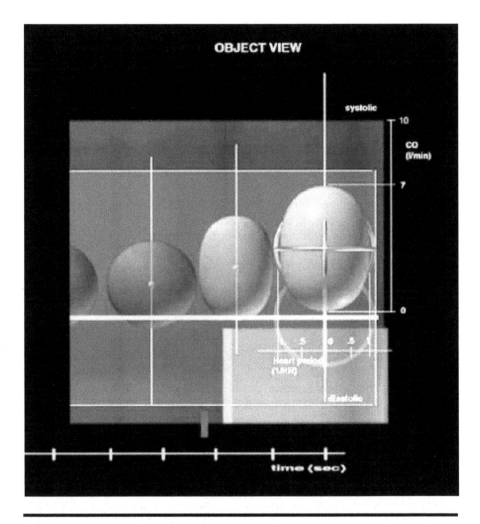

Figure 8.18   Heart object developed for a 3D display. (Reprinted with permission from Zhang, Y., Drews, F. A., Westenskow, D. R., Forest, S., Agutter, J., Bermudez, J. C., Blike, G., and Loeb, R. (2002) Effects of integrated graphical displays on situation awareness in anaesthesiology, *Cognition, Technology & Work,* **4:82–90. ©2002 by Springer-Verlag.)**

Evaluation

In a study conducted by Zhang et al. (2002), the 3D display helped the medical practitioners see changes more rapidly. In one scenario, situation awareness was higher with the 3D display compared with a traditional display. Overall, during 63% of the simulated scenarios, reliable differences were found in favor of the 3D displays.

This display (Figure 8.17) uses:

1. A trend chart of cardiac variables over time with thresholds at the left side of the display
2. Geometric features integrating cardiac variables showing state
3. Use of color and 3D to increase salience of the cardiac objects over time

Figure 8.18 uses:

1. Geometric features integrating cardiac variables showing state (heart period, cardiac output, systolic/diastolic pressures)
2. Trend chart with reference to previous cardiac states (shown by geometric shape)
3. Use of fading to make the current cardiac object salient

Sonification Display

In the third example, Watson et al. (1999) applied the principles of EID and additional analyses to design sonification displays to help anesthesiologists monitor patient state in the operating room. This is a novel and unique application of EID principles to design; for the most part, EID displays have been visually oriented.

Sonification is the representation of information as sound. In the operating room, anesthesiologists use real-time sonification to monitor current and changing patient states as they attend to other required tasks (e.g., injecting anesthetics). Typically, anesthesiologists monitor two physiological variables, heart rate and oxygen saturation, through sonification.

Sanderson et al. (2000) proposed an extension of EID to be applicable for the design of auditory displays. Their analysis required approaches in addition to WDA to come up with effective auditory displays; these approaches include the other stages of analysis of Cognitive Work Analysis (Rasmussen et al. 1994; Vicente 1999) and an "attentional mapping" stage, which maps the attentional need of the anesthesiologists. With sonification, WDA helps identify the work domain constraints and relations that need to be displayed in the auditory display. For example, patient physiological variables could map onto an auditory form. Oxygenation has been represented by pitch or tone, and heart rate has been represented by auditory pulses. Other analyses (e.g., control tasks, strategies, social organization, competencies) provide complementary insights into more effective auditory designs. These analyses were integrated to create sonification forms for anesthesiologists in the operating room.

One study on the effectiveness of these displays comes from work by Watson et al. (1999; 2000) using two-stream, five-variable sonification. Anesthesiologists attended to the sonification while performing an arithmetic task. Monitoring performance was better and interference with the primary task was lower with the sonification display compared with only a traditional visual display of patient variables. Untrained participants also participated in this study, and their performance was lower when monitoring the sonification than when information was presented visually (Watson et al. 2000).

ANALYSIS FOR DIABETES MANAGEMENT

Diabetes is a disease that has been given increased attention. The World Health Organization forecasts that more than 300 million people worldwide will have the disease by 2025. This will have a large impact on the health care systems in numerous nations. The quality of life will decrease and the burden on the health care system will increase if methods cannot be found to allow diabetics to have better control and understanding of their disease.

At the root of diabetes is the lack of insulin or the ability to use insulin properly. Insulin allows the body to absorb glucose into the cells for energy production. The main problem in diabetes management is the control of blood sugars. This is achieved through a balance of food, insulin, and exercise. If the blood sugars are not kept under control, hypoglycemia (low blood sugar), hyperglycemia (high blood sugar), and long-term complications may result. If patients can understand and manage their own disease, they will be less dependent on the health care system for management and will be less of a burden for long-term complications.

System Boundary

The work environment in this example includes the patient, the health care providers (doctor, nurse, dietician, parent, friend, and so on), food, insulin, and all the diabetic testing and administering supplies (blood glucose meter and supplies, syringes, ketone strips, and so on). The work domain includes the diabetic patient, food, and insulin. This system boundary definition is somewhat different since the patient is one of the controllers of the system.

Work Domain Analysis

The diabetic work domain was modeled at a whole body level (Figure 8.19 shows the top three levels), then in detail for balances, processes,

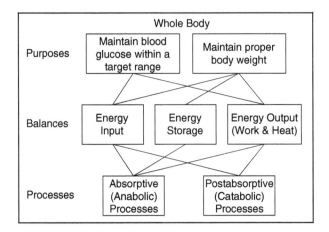

Figure 8.19 Work domain model for diabetes, whole body.

and physiology (Figure 8.20). The designer, Laura Thompson, used the same labels as Hajdukiewicz et al. (2001). Below is a description of the various levels of the AH.

Functional Purpose: There are two main purposes of the diabetic work domain: 1) to maintain the blood glucose within normal glycemic range, and 2) to maintain proper body weight. The first purpose is the dominant purpose for Type 1 diabetics, but the second purpose can be more important for Type 2 diabetics. These two purposes show the role of insulin in the balance of unusable and usable energy in the system. If there is too much energy (food) input with inadequate insulin, the blood sugars rise and hyperglycemia and ketoacidosis can occur. The glucose is spilled into the urine instead of being stored or used by the cells in the body. In this case, the insulin needs to be increased (for weight gain) or the food input needs to be decreased (for weight loss). If too much insulin is injected for the amount of food ingested, hypoglycemia results and the patient needs to ingest more sugar (glucose).

There is also the case where purpose 1 is achieved (the level of insulin matches the level of the food) but purpose 2 may or may not be met. In this case the caloric (energy) intake may be too high, too low, or adequate depending on whether weight gain, loss, or maintenance is desired. Both the insulin and the caloric intake would have to be adjusted if purpose 1 is met but not purpose 2. For many Type 2 diabetics, weight loss (purpose 2) and glycemic control (purpose 1) are very important.

Balances: This level was modeled as an energy system with energy input, storage, and output flow. The input caloric energy from the food ingested is stored in the fatty tissues and muscles, and then the energy is used by the body for normal functions and exercise. Looking at the

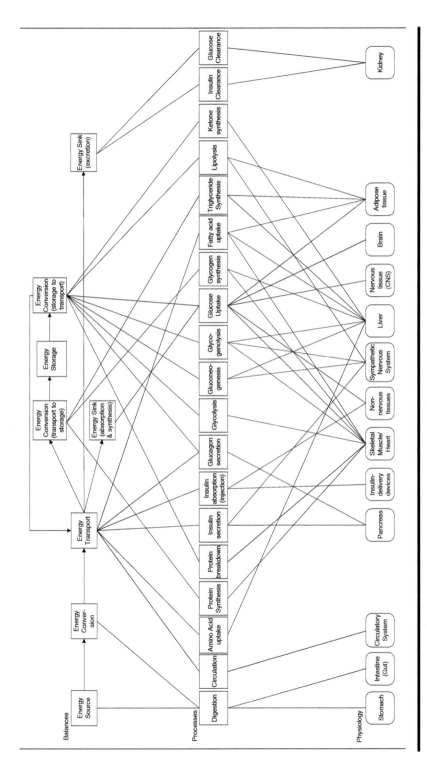

Figure 8.20 Means-ends links between the Balances, Processes and Physiology levels at the Component (Organ) level of decomposition. Only the organ components at the Physiology level are shown.

domain at this level, patients can understand how to achieve purpose 2. For example, if weight loss is needed, but patients cannot eat less food, they must exercise more to remove some of the stored energy. Also, if there is not an adequate amount of insulin, the energy is not stored or used.

Processes: This level models all the main glucose processes in the body: glucose absorption, hepatic glucose balance, breakdown/storage of fats (lypolysis/lypogenesis), peripheral glucose uptake, insulin-independent glucose uptake, and urinary glucose excretion. The processes involve glucose production, storage, utilization, and removal. This is the most complicated level to understand since it involves many interconnected processes throughout the body.

Physiology: This level describes the capabilities of the components. This level includes the major organs involved (gut, liver, brain, and kidney) as well as the circulatory system, muscles, and adipose tissue (fat). There are organ components (such as the pancreas and kidneys) and transport components (such as plasma insulin, glucagon, and glycerol). Figure 8.20 only shows the organ components, although transport components were modeled in a separate hierarchy.

Physical Form: This level describes the physical appearance, condition, and location of the components — or its anatomy. This level is not shown in Figure 8.19 or Figure 8.20. The food and insulin can be placed on this level, but it is difficult to see the physical condition of the internal organs and processes. The actual location of components is less important than their interconnections and functions.

Causal models showing the interconnections of the components were also created. From these models, the key variables in Table 8.2 were identified.

Instrumentation Availability

Table 8.3 shows some of the information requirements and their possible sources. The possible sources of information have been classified as those that can be directly measured, those that must be estimated, those that can be calculated from other variables, and those that must be simulated.

Display Design

This project is still under development at the time of writing this book. One of the goals of the project, though, has been to develop EID displays on mobile devices. The following figures show various screenshots of this application. The application was developed for multiple different device types; therefore, the screens differ in their appearance slightly.

Table 8.2   Some Variables Identified from the Work Domain Model

Variable	Units	Level
Body weight	kg or lb	Functional Purpose
Blood glucose (plasma glucose)	mmol/l or mg/dl	Functional Purpose, Physical Function
Age	Years	Functional Purpose
Energy (in and out)	Calories or Kj	Abstract Function
Glucose absorption rate (gut)	mmol/h	Generalized Function
Net hepatic glucose balance	mmol/h/kg	Generalized Function
Rate of peripheral (insulin-dependent) glucose uptake	mmol/h/kg	Generalized Function
Rate of insulin-independent glucose uptake	mmol/h/kg	Generalized Function
Urinary glucose excretion rate	mmol/h	Generalized Function
Plasma insulin	mU/l	Physical Function
Amount of injectable insulin	Units (U)	Physical Function
Type of injectable insulin		Physical Function
Time since each insulin injection	Hours	Physical Function
Action curves of the types of insulin		Physical Function
Amount of food ingested	Grams or volume	Physical Function
Type of food ingested		Physical Function
Time since food ingested	Hours	Physical Function
Unit carbohydrate (CHO) value of food ingested	Ratio of grams CHO to food amount	Physical Function
Ketone level	mmol/l or mg/dl	Physical Function
Renal threshold of glucose	mmol/l	Physical Function
Creatine clearance rate	ml/min	Physical Function
Peripheral insulin sensitivity (Sp)		Physical Function
Hepatic insulin sensitivity (S_h)		Physical Function

The work domain model was used to define information types and to structure the various displays. Figure 8.21 shows the main menu of the system. The blood glucose readings are shown immediately in terms of their general state (OK, too high, too low) and the menu shows component screens such as energy balances and body mass index (Figure 8.22).

Figure 8.23 shows screens that implement the balances level, in particular energy balances. In both cases, the screen shows the ideal energy

Table 8.3 Variables Identified from the Work Domain Model, with Availability

Measured	Estimated	Calculated	Simulated
Blood glucose	Carbohydrate	Basal energy	Plasma insulin
Urine/blood ketones	from food x	expenditure	Blood glucose
Insulin type and dose	Calories from	Body mass	Net hepatic
Height	food x	index	glucose
Weight	Calories burned		balance
Blood pressure	from exercise y		Glucose uptake
Amount of food x			Glucose
Duration and intensity			absorption
of exercise y			Renal excretion
Total cholesterol, HDL			
and LDL			
Blood triglycerides			
Microalbumin			
Creatinine clearance			
rate			

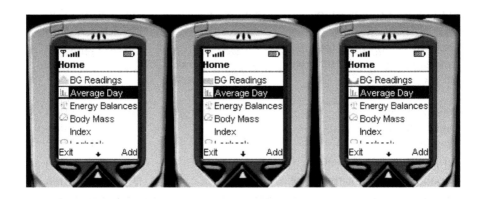

Figure 8.21 Menu for EID on a mobile device, diabetes project.

balance and the current energy balance for that day. A line attaches to the top of the graphs, adding an emergent feature to help the user tell when the energy levels are balanced. Screen 1 is in color and Screen 2 is a monochrome screen.

Figure 8.24 shows a process level screen. In this case, carbohydrate intake is shown by the bars and the curve shows the estimated plasma insulin level. The variables are combined to show the user the effects of carbohydrate intake on plasma insulin. Plasma insulin is estimated since

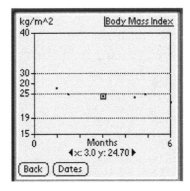

Figure 8.22   Body mass index history on a monochrome Palm.

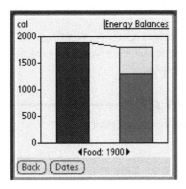
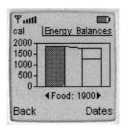

Figure 8.23   Energy Balances on color Palm and monochrome Motorola i85 cellphone.

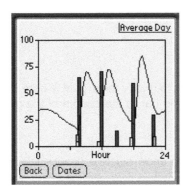
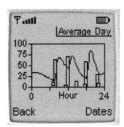

Figure 8.24   Average day on color Palm and monochrome Motorola i85 cellphone.

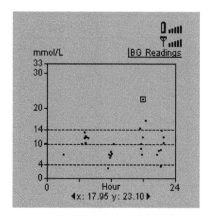

Figure 8.25   Blood glucose readings over the day on a Blackberry.

it cannot be measured directly. Blood glucose readings, in Figure 8.25, also give more information on processes as well as contributing to the Functional Purpose level status display.

▼▲▼

EYE ON VISUALIZATION

This display (Figure 8.21) uses:

1. Icons as symbols that show overall state in a menu

Figure 8.22 uses:

1. Graphs of values with thresholds
2. A box around the selected value to increase salience
3. Selected digital readout at the bottom of the graph
4. Axes that scale to show the data and thresholds as clearly as possible on this small screen

Figure 8.23 uses:

1. Stacked bar graph
2. Connecting lines between bars to give an emergent feature for balance
3. A digital readout of the selected bar at the bottom of the graph

Figure 8.24 uses:

1. Stacked bar graphs
2. Bar graphs and line graphs to show discrete (insulin injections) and continuous information (blood insulin level) on the same graph

Figure 8.25 uses:

1. A scatterplot of values over time with thresholds

▲▼▲

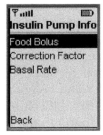

Figure 8.26 Insulin pump settings on a monochrome cellphone.

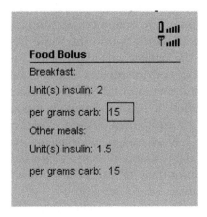

Figure 8.27 Food bolus on a Blackberry.

Finally, Figure 8.26 and Figure 8.27 show component level information. In this case, the settings of the insulin pump are shown at two levels of detail: the range of settings and the actual food bolus settings.

This implementation shows quite effectively that EID can be applied on the small screens of cellphones and PDAs, and the graphic ideas can be used as well.

Evaluation

These displays will undergo testing in the near future in a randomized controlled trial.

HANDLING THE CHALLENGES

Challenge 1: How to Choose the System Boundary

It can be particularly difficult to decide on the system boundary in medicine because the environment is very dynamic, with medical practitioners, equipment, resources, and patients continuously changing. In addition, it can be hard to determine where system boundaries lie, due to the high degree of interconnection between physiological components and the invasiveness of human intervention. The WDA framework itself, developed for process control, requires some ingenuity in order to be applied to medical systems. Some components act as control systems, again making system boundaries difficult to determine. In spite of these difficulties, there are useful aspects to the medical environment that can be modeled for the purpose of display design.

There were two dimensions to choosing a system boundary that appeared in the above examples. The first was where to bound the system in general, and the second was what to include in that boundary.

The neonatal intensive care unit and operating room examples showed two distinct system boundaries within the same larger domain. Drawing the boundary around the patient was appropriate for a patient monitoring system. To monitor the performance of specific equipment, though, the boundary needs to be drawn around that equipment. The contents of the models were quite different, with the patient model focused on physiological functions, and the equipment model focused on how that equipment worked. A designer has changing options, as well. For an effective equipment display, a tight boundary is most effective. The patient boundary would be most appropriate for integrating several pieces of equipment, or for redesigning sensors or equipment.

The diabetes example revealed another issue with system boundaries, particular to biological systems. Biological systems cannot be teased apart as easily as engineered systems. Some of the components of the system act as controllers in that system. The patients may be the work domain, but if not anesthetized, the patients also control their bodies. Our best solution so far seems to be to include these controllers and accept that control as part of the processes of the domain.

Challenge 2: How to Decide When to Stop Decomposing of the Part-Whole Analysis

Biological systems can be decomposed indefinitely in a Part-Whole Hierarchy. From the overall system, you can look at major systems, organs, tissues, cells, and subcellular structures, down to molecules and atoms. It made sense in the diabetes project to stop at the organ level and in the operating room and neonatal projects to continue to the cellular level. In these domains, cellular oxygenation could be a critical factor.

Where to stop your analysis should be discussed with subject matter experts. At a very practical level, it doesn't make sense to continue analyzing beyond the level that you can measure. This can be used as a rough guideline. If sensors beyond that level are eventually developed, your analysis can be expanded.

Challenge 3: How to Integrate Medical Information

In implementing displays from the work domain model, there can be large amounts of information that are unavailable. It becomes a challenge to integrate the available information in a meaningful way.

This could be handled in a number of ways in the design. First, the work domain model provides the context for situating these variables. This added structure provides some indication of its role in the overall determination of patient status. For example, Sharp and Helmicki (1998) organized the EID display to show the causal and structural relations associated with adequate oxygenation. Some variables associated with balances were above process variables, which support their function.

Second, the work domain model can identify which variables are meaningful to integrate and organize based on causal or AH relations. For example, the variables in Blike's display, as described by Zhang et al. (2002), were integrated based on causal models of cardiovascular function that could have been derived from a work domain model of the patient. The 3D display (Zhang et al. 2002) integrates traditionally isolated physiological data (e.g., cardiac output and heart period), and functional interactions among variables are displayed graphically. The diabetes displays have different levels of the AH on different screens, with some key-related variables available at each level.

As seen by the above examples, despite the limitations of available information, partial integration of medical information can be informed by the work domain model and may be realized.

Challenge 4: How to Deal with Issues Associated with Sensor Limitations and Availability

The medical domain is particularly constrained with sensor limitations and availability. In addition, sensors can be invasive, posing significant risks to the patient. Sharp and Helmicki (1998) noted four categories of mapping in terms of availability of individual sensors and their use: direct mapping, analytical mapping, heuristic mapping, and no mapping. With direct mapping, the identified variables are sensed directly and values used. With analytical mapping, the identified variables required the use of an existing analytical model to derive values. With heuristic mapping, the identified variables were derived using a heuristic map. These relationships between observable and unobservable variables were qualitative and subjective in nature. With no mapping, the identified variables had no mapping between observed quantities and the target variables.

Hajdukiewicz et al. (2001) provide four different categories in terms of evidence used across sensors: one-to-one mapping, convergent mapping, divergent mapping, and no mapping. With a one-to-one mapping, one sensor maps onto one patient variable. For example, checking a patient's pulse provides information about heart rate. With convergent mapping (or redundancy), many sensors map onto one patient variable. Practitioners use this method to reduce the high level of uncertainty in measurements from the environment (e.g., artifact, noise, and calibration errors). For example, heart rate can be determined directly from the ECG signal as well as indirectly from other monitor signals (e.g., arterial blood pressure waveforms). With divergent mapping, some sensors provide evidence for many patient variables. For example, the ECG waveform provides evidence for heart rate, heart rhythm, and adequate myocardial oxygenation, among others. Finally, with no mapping, some sensors do not map onto any patient variables. The pressure in an unused oxygen tank at the anesthesia workstation is an example.

In the diabetes example, the possible sources of information were classified as those that can be directly measured, those that must be estimated, those that can be calculated from other variables, and those that must be simulated.

Despite the limitations of sensors and their availability, the work domain model can provide a way to organize information by providing contextual information about current medical knowledge. This knowledge is constantly evolving as medical innovations and discoveries are made. In addition, the work domain model can also provide a way to strategically place sensors and focus development efforts in sensors that will have an impact on patient monitoring and interface design.

Challenge 5: How to Determine Information Requirements for Different Roles in the Medical Environment

The work domain models from the operating room example demonstrate how roles can influence information needs. For example, the information requirements for surgeons are different compared with anesthesiologists, although both need the same information from overlapping regions of the patient work domain model (Figure 8.15 and Figure 8.28). Once these requirements are determined for each person or team, an appropriate form may be designed that is compatible with the capabilities of the medical personnel (e.g., using perceptual features instead of numbers, requiring less analytical reasoning). For example, as shown in Figure 8.28, surgeons may primarily need a view of the surgical field and an anatomical map. Anesthesiologists, on the other hand, may need integrated higher level representations of how well the patient is coping throughout the

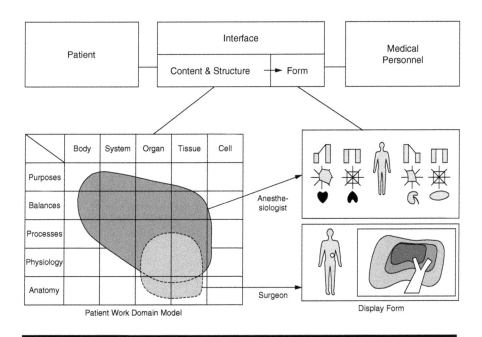

Figure 8.28   Designing for different roles. (Reprinted from Hajdukiewicz, J. R., Vicente, K. J., Doyle, D. J., Milgram, P., and Burns, C. M. (2001) Modeling a medical environment: An ontology for integrated medical informatics design, *International Journal of Medical Informatics* 62: 79–99. ©2001 with permission from Elsevier.)

surgical procedure. This is indicated in Figure 8.28 as a series of integrated displays.

The displays for the neonatal intensive care unit example were designed for specific use by the medical practitioner (e.g., physician or nurse) responsible for care of the newborn. The displays noted in the operating room example focused on information specific to higher level physiological function, thus supporting the anesthesiologist's role. Finally, the displays in the diabetes management example were designed for care provider and patient use.

Challenge 6: How to Extend EID to Other Sensory Modes (e.g., Auditory)

From the work of Watson et al. (1999; 2000), it was demonstrated that EID could be applied to other sensory modalities besides the visual channel. A work domain model defines the information requirements independent of form. Sanderson et al. (2000) note that for effective auditory, vestibular, haptic, and olfactory design, knowledge will be required about when a control task needs to be supported in focal or nonfocal attention and how many channels are required to convey the environment to the operator.

A Haptic EID?
Stoner, Wiese and Lee (2003) have used their analysis (discussed in Chapter 3) to propose a haptic ecological interface for driving. This figure shows how safe regions of travel (adopted from Gibson and Crooks (1938) could be conveyed through vibrating areas embedded in the driver's seat.

9

SOCIAL SYSTEMS

Social systems are not particularly product oriented. They involve the exchange of goods or the delivery of experiences. In designing for these systems, the quality of the experience or the service is a key factor in having people return to these systems. Stores, websites, and entertainment systems would fit into this category.

In many cases, EID is probably not the best design framework to use for these systems. Designs for these systems need to focus on ease of use and the user experience. Occasionally EID can play a role, though. EID is a good approach to add to your design process if your system needs to explain or clarify a concept to your users. You may want to use EID selectively in this kind of an approach.

In this chapter we look at the use of EID for a video poker game. In particular, EID is being used to display to the user the workings of the game. The display is designed to help educate problem gamblers and guide them towards making better decisions. In this way, using EID was a good fit for this design problem. Your average casino, though, is certainly not designed on EID principles, and quite deliberately. Some of the fun and entertainment comes from not always knowing how the system works.

CHALLENGES

Social systems have several unique challenges. First, they don't necessarily follow the physical principles of other systems. Social systems are driven on social values like money, service, and the quality of the experience. The user is not just a controller in the system, the user participates in the system. In a sense the user is also a component. If the user is not active in shopping, gambling, or participating in the system, the system doesn't run as designed. This creates two different interacting domains: the user's domain and the provider's domain. Third, the design must create a certain user experience such as entertainment, exploration or discovery.

CHALLENGES WITH SOCIAL SYSTEMS

Modeling money and value
Modeling two different but tightly connected domains
Developing entertaining visualizations

CASINO GAMBLING

Problem gambling, and in more extreme cases, pathological or compulsive gambling, is defined as a progressive addiction "characterized by increasing preoccupation with gambling, a need to bet more money more frequently, restlessness or irritability when attempting to stop, 'chasing' losses, and loss of control manifested by continuation of the gambling behavior in spite of mounting, serious, negative consequences" (Problem Gambling Research Group 1999). It is estimated that as many as 5.5 million Americans may be problem gamblers and as many as 148 million may be at some risk of showing problematic gambling behavior (National Gambling Impact Study Commission 1999). In Canada, the number of problem gamblers is estimated between 600,000 and 1.2 million (National Council of Welfare 1996). A key factor in becoming a problem gambler is access to gambling opportunities. The number of gambling opportunities is certainly rising. In Canada, many new casinos have been developed in cities along the Canada–U.S. border, hoping to attract patrons from both countries; similar developments are occurring in the United States (Searcey 1999). Another recent development is the prevalence of online casinos available through the Internet. These online gaming sites bring gambling opportunities directly into people's homes.

For display design, gambling presented an interesting problem. While we cannot influence the physiology of addiction, some aspects of problem gambling are related to poor decision-making behavior. Problem gamblers lose sight of the negative consequences of their actions and are overly optimistic about their chances of winning back their losses. From a cognitive perspective, this shows a poor mental model of the gambling environment and a lack of understanding of how randomness and odds really work. This lack of understanding results in, effectively, inappropriate and ineffective action being taken by the operators. Problem gamblers also play longer and lose more money than other gamblers (National Council of Welfare 1996), possibly forgetting how much they have lost and how long they have played. The development of computerized and online gaming systems offers an opportunity for display design to be used as a lever to influence this problem. A well-designed display could keep track of time spent and losses, both memory tasks, as well as revealing how the gambling environment works.

System Boundary

We drew the system boundary to include the patron and the house or gambling organization. We included both in the system so that we could capture the interactions between them. This boundary is somewhat different than in the other systems we have looked at. In the other systems we have deliberately not included the operator in the boundary. The reason we included the patron within the boundary this time is that the patron is something that we want to control. The perspective taken in this project is a level higher than in the other examples. In this view, the patron is an element within the system and so is included in the model.

Work Domain Analysis

Purposes

The patron and the house have very different objectives in the gambling domain. The house is interested in making a profit, whereas the patron may be seeking entertainment and winning; 90% of problem gamblers say they gamble for entertainment, and 84% say they gamble to win money (National Council of Welfare 1996). We identified these two conflicting purposes between the patron and the house as the first step in our model.

Because we had two conflicting purposes and two distinct domain elements, patron and house, this led to a two-domain model split across all five levels (Figure 9.1).

Abstract Function

To describe Abstract Function for gambling we had to determine the fundamental principles driving gambling. One way to determine principles is to look for things that flow and move. Therefore, we described the gambling domain as being driven by the flow of money, or more abstractly, value relationships. The house is set up so that more money flows into the house than leaves, so that it attains its purpose of making a profit. The money that flows in must come from the gambler or patron. Some of the money flows out to the patron as winnings, but ultimately not as much as is retained by the house. The gambler also extracts value from the exchange in the form of entertainment, possibly expectations of winnings. However, problem gamblers play longer than other players,

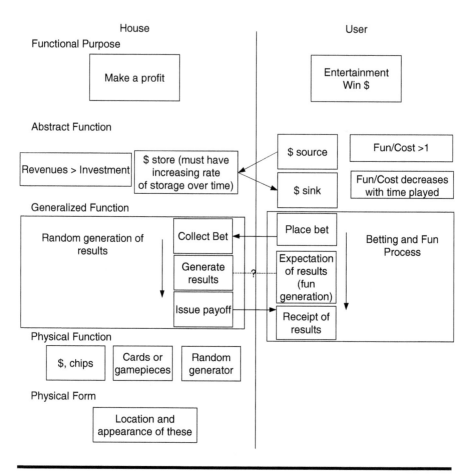

Figure 9.1   Work domain model for casino gambling.

and losses typically increase with time played. We modeled this as a "fun/cost" ratio decreasing over time played, which is ideally greater than 1 but decreases as a patron moves into a problem gambling situation.

Generalized Function

There is one main process that involves both the house and the patron. The house accepts money in (the "bet"), generates a random result through different ways depending on the casino game, then issues a payoff based on certain odds. The odds are set to meet the house's purpose of making a profit and are a key constraint in understanding the system. The patron takes the money out, decides whether to make another bet, and if so puts money back into the process. We identified the degree of fit between the patron's decision making and the reality of the random generation, probability, and odds-setting to be a key area to improve in our display design.

Physical Function

The main components of our system were money and value-related gambling objects (cards, chips). This would include virtual value objects in an Internet gambling situation.

Physical Form

This would describe the appearance of money, cards, chips, and so on and their locations (physical or simulated).

Display Design

Keeping in mind that our design goal was to improve the mental models of problem gamblers, our work domain model indicated three key areas for display design. The first thing we wanted to show in our design was that the house always makes a profit. This came from understanding the purpose of the house and from modeling the flow of money in the gambling situation. We wanted to show that the house achieves this profit by setting odds in its favor, the key part of the process identified at the Generalized Function level. In the problem gambling situation, the entertainment value decreases with increased play, whereas the chances of losing increase with increased play (AF, patron side). This is counter to the patron's goals (FP, patron side).

We limited our design to the game of video poker and applied the design ideas that we obtained from our work domain model. To give a richer sense of the design process we have included preimplementation sketches of key visualizations. The following subsections describe the design.

Connecting Losses and Time Played

Two aspects of our display focus on showing the loss of money or value with time played by the patron. On starting up the display, the patron is asked to answer "How much money are you willing to lose?" and to set a time limit for play using a "downward spiral" visualization (Figure 9.2). This visualization suggests that longer play is more problematic and shows that losses increase with time played (AF). The visualization with the spiral introduced the icon of a dog that became a metaphor throughout the display. We also included a trend display (along the bottom of Figure 9.4)

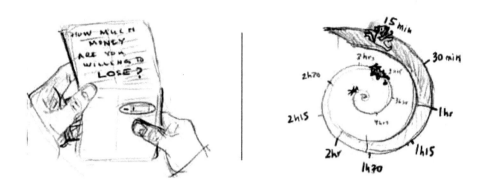

Figure 9.2   Design sketch of time–money relationship representations.

showing winnings and losses over the time played. The frame of reference on the trend display is money vs. time, and colored bars in the background show the current earnings or losses against contextual regions of green for profit, orange for losses within predetermined limits, and red for losses beyond the predetermined limits. We also showed bets and potential paybacks directly as money rather than chips, to emphasize the money flows of the domain.

The Dirty Dog Metaphor

We modified a visualization by Kuk et al. (described in Vicente 1998) from a project called the "PowerPig." The PowerPig software used the metaphor of a pig, changing weight levels from lean to rotund, to reflect a user's electrical power consumption. We modified this visualization to create a dog, changing from a suave "cool dog" to a "dirty dog" rummaging through trash when the gambler has exceeded his allowable losses, thereby showing the overall state of our gambler, implementing our FP level (Figure 9.3). This is shown in the "cool dog" state in the top left corner of the overall display (Figure 9.4). We referred to this as a "behavioral representation," as it shows the behavior of the patron.

Representation of the House–Patron Relationship

It was critical to show the patron that the house always wins and odds are deliberately set to favor the house. We used the metaphor of an unfair race to demonstrate this idea (shown in the top of Figure 9.4). Depending on the odds of the gamble that is being considered, the house is shown at different levels of advantage, indicated by the technology that the house uses in the race. Regardless of the situation, the house is always shown as advantaged over the patron.

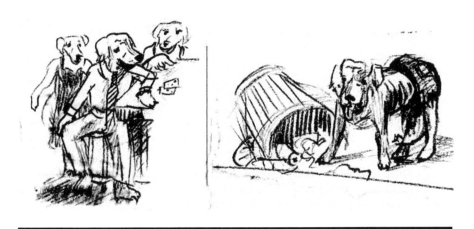

Figure 9.3   The dirty dog metaphor.

Behavioural
Representation

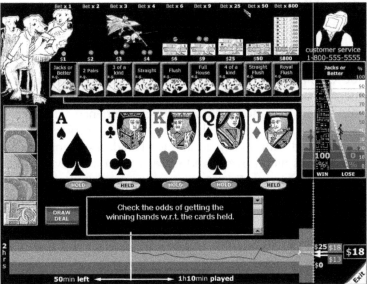

Visualisation
of the Odds

Gambling History

Figure 9.4   Overall display (main screen). (Reprinted with permission from Burns, C.M., and Proulx, P. (2002) Solving social problems through interface design. *Ergonomics in Design,* 10(4), 10–16. ©2002 by the Human Factors and Ergonomics Society. All rights reserved.)

Eye on Visualization

This display (Figure 9.4) uses:

1. A dog that changes state as an overview display in the top left corner
2. A trend chart of money over time with thresholds at the bottom of the display
3. Connected bar graphs to show odds, with an emergent feature to show the likelihood of winning (center right side)

Odds Visualization

We showed the odds of winning using an elaborate bar graph (right-hand side of Figure 9.4), with one bar for the chances of winning and one for the chances of losing. The tops of the bars are connected to create a more powerful visualization. A person is shown, balancing on the bar, having an easy time sliding down the bar when the chances of winning are good, and a hard time climbing up the bar when the chances of winning are poor and losing quite good.

Evaluation

A usability test was also conducted followed by a review with two domain experts who develop problem gambling software. In general, most respondents confirmed that the information provided in the display would be very useful and would enhance their gambling experience. The review with the domain experts did not reveal as many usability problems as the usability test, conducted with undergrad and grad students in engineering. Evaluating EID displays with novices and experts will often reveal differences between the two groups. Expert users may recognize the principles that underlie the graphics more immediately. Novice users may require training or longer experiences with the display to understand the novel graphics.

In this case, the experts mentioned that the spiral on the time limit mechanism made for a good analogy, very suggestive of what might happen if a gambler plays for too long. One of the two experts knew right away how to extract the meaning of the configural display for the odds. The other expert did not realize at first the difference between the platform and the stairs. They also said that the dog representation was a good idea. However, results from the evaluations on the behavior-shaping constraints and the other representations at the top of the gaming workspace are not as valid as the results on the other sections, since these sections were not nearly implemented to the same degree. Both experts concluded that the display would be very useful and very appealing to problem gamblers. The gambling history view was considered particularly useful.

HANDLING THE CHALLENGES

Challenge 1: Modeling Money and Value

Money can be modeled similarly to other physical entities. It can move and be added and subtracted. Money gets exchanged for goods or services, though, and this is where the model becomes more challenging. The exchange of money for goods is dependent on different levels of value. The value assessments between the user and the provider must be different, and indeed higher by the user, for the transaction to occur. So when we are balancing money, we are really discussing the flow of value. In your model you will often have to include a description of what is driving value in the current system.

In the casino gambling example we modeled the basic flow of money from the user to the house. It is an unbalanced flow, since the house acts to retain or store some of the money as its profit. The user drives the exchange of money. The user must perceive value in the experience to continue the flow. This value either comes from the entertainment of the experience, or from expectations of winning. The expectation of winning reflects a fundamental misunderstanding of how the house part of the domain works.

Challenge 2: Modeling Two Different but Tightly Connected Domains

In this example there were two different parties: the user and the house. These parties had two completely different objectives. This led to the two-domain split model that was used. In many ways, this system is similar to the military frigate example. Again, the frigate had three different domains with conflicting purposes, and so a three-part model was used. The clues to determining how many models should be used are:

- Can the parts of the domain be physically isolated from each other and exist without each other?
- Do the parts of the domain have different objectives?

Challenge 3: Developing Entertaining Visualizations

The third challenge of this project was to implement ecological visualizations in a way that would be meaningful and attractive to a gambling patron. Bar graphs and polar star displays were not considered to be good options for this kind of entertainment-driven system.

The basic rules of display design still had to be used. We needed to show the current values and their context. The dirty dog display is a metaphor equivalent to the polar star. It is a status display that integrates several different variables. The odds bargraph is a modification of a basic bar graph with a balance line. The background and the addition of the figure have enhanced the display. The race is an attempt to show current values using the race as a metaphor that creates context. The general display concepts have been maintained but implemented in a more attractive format.

10

USING EID WITH OTHER METHODS

EID is one of many approaches available to designers for human–computer interaction in complex work environments. Other popular approaches include Cognitive Work Analysis, Task Analysis, Situation Awareness Analysis, Contextual Inquiry, Use Case Scenario Design, GOMS (Goal, Operators, Methods, and Selection Rules), Participatory Design, User Interface Principles, and Usability Evaluation. Some of the outputs of these analyses are alternatives to that of EID for informing interface design. For example, Contextual Inquiry provides some alternatives to EID; both methods have work models based on the constraints of the work environment. Other approaches provide complementary perspectives that are useful in filling gaps to interface design. For example, EID identifies how component resources map functionally to purposes, while Task Analysis identifies activities that enable a user to achieve particular goals.

This chapter begins to shed light on the specific work environment and interface lifecycle areas EID is intended to target and where it sits in comparison with the methods mentioned above. This will provide some insight into how EID can be used with other methods to bridge gaps and achieve effective final designs. First, a frame of reference is introduced that attempts to bound relevant aspects of the work environment and interface lifecycle space, within which methods could be roughly situated. Second, each method is briefly described with examples. Third, each method is roughly mapped onto the framework and generally compared with EID. This chapter is not intended to include detailed evaluations or critiques of various methods, noting the benefits and limitations of each; that in itself would take at least a book or two to discuss. Each of the chosen methods has a unique and relevant contribution to interface design; there are also many other methods not discussed here that are relevant to the discussion.

The mapping is meant to provoke thought by analysts and designers in how they may incorporate and integrate various tools for interface design.

This chapter points out a couple of facts worth noting: EID is not a panacea for interface design, but it can help to solve focused aspects of the interface design problem for complex work environments, and other methods can help fill in the gaps for effective interface design.

A FRAMEWORK FOR DEFINING OPPORTUNITIES TO USE EID WITH OTHER METHODS

To assess opportunities in leveraging other methods with EID, a high-level framework was created that provides a coarse map of the generic problem space covering both the work environment or context (i.e., the work domain of focus, activities, and capabilities of people and technology) and the typical lifecycle process used for interface design (i.e., information requirements, interface design, and evaluation).

Work Environment

The work environment for the purposes of this book is defined as the domain where work needs to be performed, the activities (both physical and cognitive) that are required to achieve goals and the domain purpose, and the characteristics of people, culture, and technology (e.g., capabilities) that allow them to perform activities within the work environment. Aspects of this dimension are taken from numerous sources (e.g., Vicente 1999; Rasmussen et al. 1994; Beyer and Holtzblatt 1998).

As shown in Figure 10.1, there are three main parts to the work environment. The work domain describes *why* and *where* work needs to be done, the components and resources available for fulfilling the domain purposes, the functions of these resources, and the principles, relationships, and constraints that govern domain behavior. Activities include the tasks and procedures that need to be performed to achieve particular goals, the strategies for performing these tasks, troubleshooting, decision making, and coordination. Activities can be considered as the *what, when*, and *how* associated with action in the work environment. People and technology define the capabilities, competencies, organization, role allocation, responsibilities, communication, and culture in the work environment. These characteristics can define the *who* when considering the work environment.

The work environment is a relevant dimension in situating the above-mentioned approaches because certain design approaches cover many aspects of the work context at a broad or a focused, deep level. Thus, an understanding of how the approaches map onto this frame of reference would be useful for situating EID.

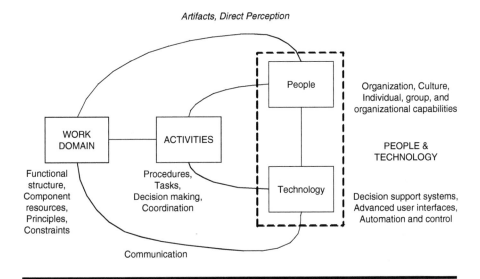

Figure 10.1 Defining the work environment.

Work Domain

The work domain is a focused part of the work environment, typically including the equipment resources that align with the purposes for which the work environment was created (Vicente 1999; Rasmussen et al. 1994). For example, in a petrochemical plant, the work domain could be defined as all the equipment and functions in a unit that exist to meet operating requirements within a designed range. Work domains can be simple or complex. For the purposes of this book, we generally focused on complex work domains, defined as distributed, with multiple degrees of freedom (Rasmussen et al. 1994; Vicente 1999). A complex work domain typically requires monitoring and control by an individual or operations team. Different types of constraints govern the work domain dynamics. The work domain includes a physical and functional representation on the purpose-related aspects of the work environment.

Activities

Activities describe what needs to be done, how and when it can be done, and where it is done (Vicente 1999; Leplat 1991; Rasmussen et al. 1994; Nardi 1996). These can include troubleshooting, problem solving, decision making, executing procedures, and performing actions. Activities are the link to the environment and the facilitator or actor.

People and Technology

This part describes how people and technology are allocated to different aspects of the work domain and activities, and their associated capabilities and limitations across experience levels. It is a mapping of people, technology, artifacts, and tools to meet the requirements of the work domain and activities that need to be performed (e.g., communication). In addition, it includes the social organization and communication structures and influences of actors.

Interface Lifecycle

Another relevant dimension when situating EID to other approaches for interface design is coverage with elements of the interface lifecycle process itself. For the purposes of this discussion, interface design is typically a component of the general software design process when deploying or maintaining systems in complex work environments. As shown in Figure 10.2, the general approach to interface design takes on three stages in its lifecycle: information requirements generation, interface design, and evaluation. In practice, these stages are not performed in sequence and are iterative throughout the lifecycle of the design.

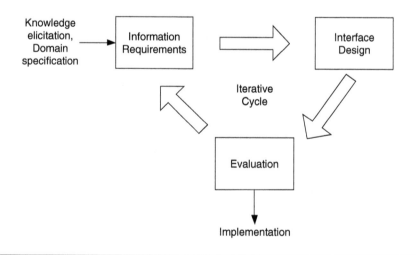

Figure 10.2 Generic iterative process for the interface lifecycle.

Information Requirements Generation

The information requirements generation stage is defined by eliciting knowledge from experts, manuals, textbooks, experiments, and heuristic rules, and by analyzing the domain, activities, and characteristics of people and technology. The outputs of this stage are the content requirements for interface design and can include variables, relationships, constraints, dependencies, goals, states, and organization. Some approaches, such as Task Analysis, specifically are focused on information requirements in terms of content, and rely on other approaches to fill in the gaps for actual interface design and evaluation.

Interface Design

The interface design stage involves taking the information requirements that define the content of what should be displayed, and creating appropriate forms based on conditions of use, including constraints and dependencies associated with the activities that need to be performed, and the people or technology performing the activity. The outputs of this stage are actual designs that map information requirements directly to interface forms. Chapters 3 and 4 discuss examples of the mapping of requirements to interface forms.

Evaluation

The third stage in the design lifecycle process is evaluation. After the interface has been designed, evaluation of its effectiveness with the target user group and stakeholders is essential. Evaluation can take on various forms from expert heuristic reviews, user group interviews, and paper prototypes to low-fidelity and high-fidelity simulations. Depending on the phase and goals of evaluations, certain approaches are more feasible and desirable than others. Usability Evaluation, discussed in a subsequent section, provides some of the tools and methods for interface evaluation.

Design Framework

Combining the two dimensions, Figure 10.3 shows the resulting matrix, with each cell presenting what content would be included. The design problem will define which cells are relevant and which methods could provide the most impact on an effective design. For example, a domain such as a call-center application is activity-intensive and somewhat predictable with relatively few work domain constraints compared with activity constraints; an approach that focuses on activities is more suitable compared

Work Environment

		WORK DOMAIN	ACTIVITIES	PEOPLE & TECHNOLOGY
	INFORMATION REQUIREMENTS	Content specific to the constraints, limits, relations, principles, components, and physical resources of the work environment.	Content that is directly related to tasks, sequences, dependencies, decision making processes, strategies, execution constraints, detection and diagnosis.	Content associated with the allocation of roles, social organization, competencies, skills, culture, and technology that are or need to be in place for the system to function properly.
	INTERFACE DESIGN	Design forms that are derived from information requirements and directly related to the constraints and relations of the work domain.	Design forms that display task-specific formats to support the detection, diagnosis, decision making, and execution of tasks.	Design elements that facilitate coordination across the organization and convey roles and responsibilities of different actors.
	EVALUATION	Evaluation methods that verify the appropriateness of interface design formats associated with the work domain based on the conditions of use.	Evaluation methods that verify the appropriate interface design formats of conveyed activities, dependencies, goals, and states, for the conditions of use.	Evaluation methods that verify the appropriate interface design formats assigned to different actors, and communication mechanisms of organizations for the conditions of use.

Interface Lifecycle (vertical label on left)

Figure 10.3   A framework for situating methods that is relevant to interface design.

with one that primarily focuses on the work domain. In contrast, a domain that is flexible, but highly constrained by science and the laws of physics, requires an approach that rigorously models the work domain.

In the sections below, we discuss how the framework can be used to generally situate the methods with each other in terms of higher level coverage. Some methods are concentrated in one or two cells; other methods are broad. Concentration and breadth are shown by coverage associated with populated cells. Shaded cells indicate that the method can be used in part to satisfy the requirements and are driven by a direct focus, theory-based detailed models or principles, heuristics, or empirically based techniques; blank cells convey that the method does not primarily contribute to requirements noted in the cell and may or may not have an indirect influence. There are advantages and disadvantages to any method. When designing interfaces, it is important to know how to effectively combine these tools to get the most out of the activities engaged.

There are conceivably other frameworks and dimensions that may be useful for comparison (e.g., time required to use the approach, cost/benefit, and so on); for our purposes, the framework presented serves as an effective, coarse approach for situating EID within the context of other methods.

EID AND OTHER METHODS

The following methods were chosen for discussion on how they could fit with EID: Cognitive Work Analysis, Task Analysis, Situation Awareness Analysis, Contextual Inquiry, GOMS, Use Case Scenario Design, Participatory Design, User Interface Principles, and Usability Evaluation. In each case, we present a brief description, relevant tools, an example, mapping with the framework, and a discussion. We have attempted to highlight the key features of each approach, but have not described the methods in elaborate detail; however, key references have been provided for further reading. In reality, some of these methods have blended together, utilizing each others' tools as appropriate. For example, EID has been used with Task Analysis to inform design, and with Usability Evaluation to assess and iterate on the designs.

The main example used throughout for convenience and continuity is DURESS II as previously described. We make the assumption, hypothetically, that two people need to coordinate DURESS II from different locations and that level control automation exists in one of the tanks; this will allow us to discuss the perspective of people and technology.

From this analysis, you will see that EID provides a set of tools that can be useful in helping develop advanced displays and solutions. Most likely, it will not be the only method that is required to develop an effective design. Integration with other methods that provide different insights and complementary perspectives will help achieve more effective solutions. We discuss these comparisons and integration opportunities in this chapter to help designers understand where EID and other methods fit in for effective display design.

Ecological Interface Design

EID has thoroughly been discussed in previous sections (see Chapter 2 through Chapter 4). In this section, we situate EID using the framework shown in Figure 10.3. Other methods mentioned will be compared with EID using this framework.

The work domain model for DURESS II is shown in Figure 10.4; the circled variables were used to create the EID interface (shown in Figure 10.5). This interface was developed by mapping the variables from the work domain model onto the display.

Mapping onto the Framework

As shown in Figure 10.6, EID mainly focuses on information requirements and interface design, focusing on work domain and human capability

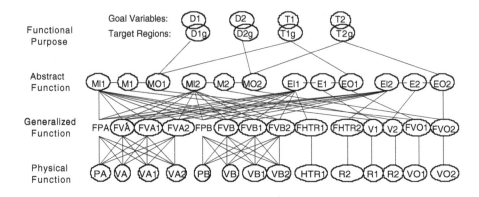

Figure 10.4 Work domain state information available on the EID interface of DURESS II. (Reprinted with permission from Hajdukiewicz, J. R. (2001) Adapting to change in complex worlds: A study of higher-level control and key success factors in a process control microworld, unpublished thesis, University of Toronto.)

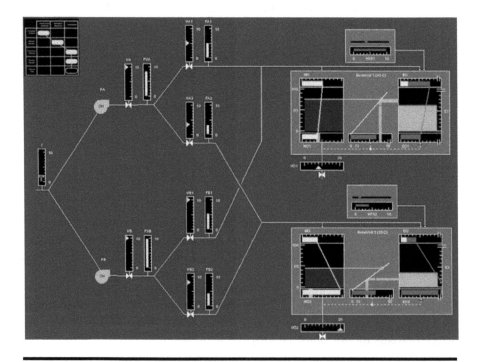

Figure 10.5 EID interface for DURESS II displays all levels of information from the work domain model. (The top left graphic indicates that most of the information at that level of the work domain model is displayed.) Adapted from Pawlak, W. S., and Vicente, K. J. (1996) "Inducing effective operator control through ecological interface design," *International Journal of Human–Computer Studies*, 44:653–688.

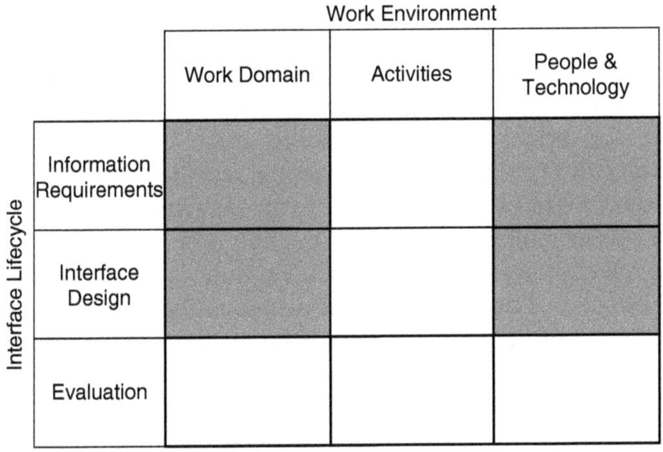
Work Environment
Work Domain
Activities
People & Technology
Interface Lifecycle
Information Requirements
Interface Design
Evaluation

Figure 10.6 Mapping of EID onto the design framework.

elements. EID captures the purpose-related constraints and relations, and considers the capabilities of people when creating designs. These requirements are obtained with reference to primarily the perceptual capabilities of people. By conducting a WDA, the designer has insights into the constraint requirements associated with design forms that link to the functioning of the work domain.

Cognitive Work Analysis

Rasmussen et al. (1994) developed the Cognitive Work Analysis framework using the constraint-based approach for analyzing complex socio-technical systems for adaptive performance. This framework identifies five conceptual distinctions or phases of analysis, representing different aspects of work for complex socio-technical systems: work domain, control tasks, strategies, social organization and cooperation, and competencies. The following discussion of Cognitive Work Analysis is based on that provided by Vicente (1999):

- *Work Domain:* This analysis results in a representation of the system being controlled, independent of any particular worker, automation, event, task, goal, or interface. A work domain analysis is like a map in that it shows the lay of the land independently of any particular activity on that land. That is, it shows the possibilities for action. The Abstraction Hierarchy is a primary tool to model the work domain.
- *Control Tasks:* This analysis includes the goals that need to be achieved, independently of how they are to be achieved or by whom.

This analysis identifies the constraints that govern activity on the work domain (as opposed to the constraints that govern the work domain itself). In other words, the focus is on identifying what needs to done, independently of the strategy (how) or actor (who). The decision ladder is a primary tool to model the control tasks.

- *Strategies:* This analysis includes the strategies by which particular control tasks or work processes can be achieved, independently of who is executing them. They describe how work process goals can be effectively achieved, independently of any particular actors. The information flow map is a tool to model the strategies.
- *Social Organization and Cooperation:* This analysis deals with the relationships between stakeholders and any automated systems. This representation describes how responsibility for different areas of the domain may be allocated to stakeholders, how control tasks and work processes may be allocated among stakeholders, and how strategies may be distributed across stakeholders. Thus, a social organizational analysis describes how stakeholders may be organized into groups or teams, how they may communicate and cooperate with each other, and what authority relationships may govern their cooperation. This analysis also includes the culture of the social organizations involved.
- *Competencies:* This analysis represents the set of constraints associated with the elders and workers themselves. In addition to considering generic human capabilities and limitations, this analysis also identifies the particular competencies that various workers should exhibit if they are to function effectively in their work environment. Different jobs require different competencies. Thus, it is important to identify the knowledge, rules, and skills that workers should have to fulfill particular roles in the organization effectively. The skills, rules, and knowledge framework is a primary tool to model competencies.

The different analyses provide the basis for identifying constraint boundaries in the work environment that shape behavior. Each distinction has a set of modeling tools to assist in identifying these constraints. Vicente (1999) and Rasmussen et al. (1994) provide detailed discussions of how these modeling tools can be incorporated in the analysis.

Example

Example outputs of two analyses are presented for DURESS II: Control Task and Strategies Analysis. Figure 10.4 shows an example of the outputs

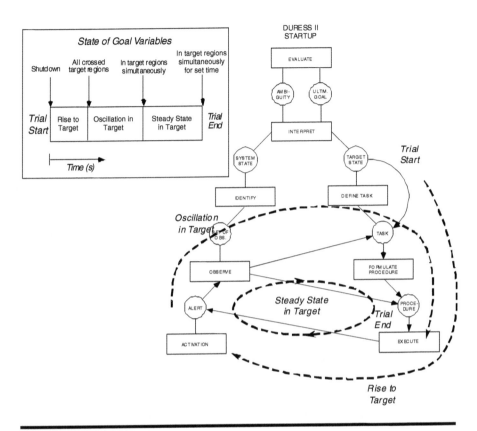

Figure 10.7   An example of the outputs of a Control Task Analysis for DURESS II startup.

of a WDA for DURESS II. More details on these analyses are provided in Vicente (1999).

Figure 10.7 shows an example of the outputs of the Control Task Analysis using the decision ladder. The decision ladder helps structure action dependencies in decision making and performing tasks. The startup process includes three general stages: rise to target, stabilization, and steady state. The decision ladder provides the structure for activities that need to be performed during this task. Initially, DURESS II is shut down. In the first stage, the operator is required to turn on the process. This includes identifying the target state, identifying the task, formulating and identifying a procedure, and executing a set of actions to start to move the goal variables to their target states. In the second stage, the operator is required to stabilize the goal variables around their target states; this may require the reformation of tasks and procedures or adjustment of actions. In the final stage, the operator is required to keep the goal variables in their target regions; this may require the adjustment of actions.

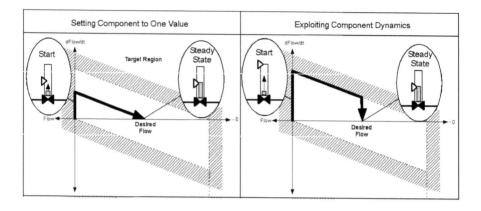

Figure 10.8 **Local strategy alternatives for moving a goal variable to its target state.**

An example of the strategies analysis is the way operators can utilize components to reach the target states. Figure 10.8 shows two alternative strategies in achieving a local target state based on the interaction between component (i.e., a valve) and flow dynamics. The graphs show the rate of change of flow, as compared with flow for two conditions:

1. Setting Component to One Value (left graphic in Figure 10.8): The operator sets the component setting to the target flow rate. In this case, the operator only has to make one control move.
2. Exploiting Component Dynamics (right graphic in Figure 10.8): The operator exploits the component dynamics by increasing the rate of change initially, then moves the component setting to the target setting when the target state is reached. In this case, the operator needs to make at least two control moves, but the time to read the target state is shorter compared with Condition 1 because the operator is exploiting component dynamics.

Discussion

Cognitive Work Analysis has some tools that are also used with EID (e.g., the Abstraction Hierarchy). However, Cognitive Work Analysis is a more comprehensive method in certain respects, because it tries to fill in gaps associated with activity and action dependencies, and social organization and cooperation constraints. EID is also more focused on providing direct links to interface design; Cognitive Work Analysis is more generic and can provide insights to other areas of concern (e.g., training requirements).

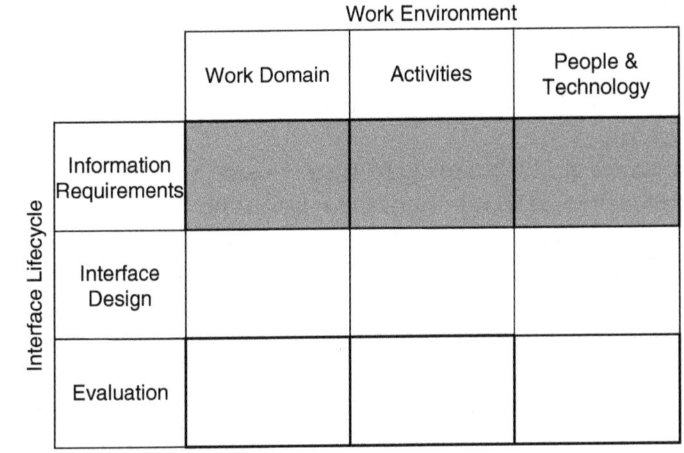

Work Environment

	Work Domain	Activities	People & Technology
Information Requirements			
Interface Design			
Evaluation			

Interface Lifecycle

Figure 10.9   Mapping of Cognitive Work Analysis onto the design framework.

Mapping onto the Framework

Cognitive Work Analysis has a number of tools for information requirements generation that map to all aspects of the work environment, as defined previously. As shown in Figure 10.9, WDA maps directly onto the Work Domain. Control Tasks and Strategies map onto Activities. Competencies map onto People and Technology. Finally, Social Organization and Cooperation map onto all three.

Task Analysis

At the core of Task Analysis is the modeling of user or actor actions in the context of goals to be met. There are various forms of Task Analysis that have appeared in the literature. Some of the more common ones include Sequential Task Analysis, Timeline Task Analysis, Hierarchical Task Analysis, Cognitive Task Analysis, and Control Task Analysis (previously discussed). In general, actions are analyzed to move from a current state to a target or goal state. The approaches have different perspectives on how to define actions: explicitly or implicitly.

- Sequential Task Analysis is a systematic process for organizing activities and can be described in terms of the perceptual, cognitive, and manual behavior required of the user (Kirwan and Ainsworth 1992). The analysis may show the sequential and simultaneous manual and intellectual activities of a person, as well as interaction with equipment.

- Timeline Task Analysis is an approach that helps assess the temporal demands of the tasks or sequences of actions and compare them to the time available for the execution of the tasks (Kirwan and Ainsworth 1992). Sequential relationships among the tasks can be analyzed.
- Hierarchical Task Analysis is a broad and more sophisticated approach used to represent the relationship between what tasks and subtasks need to be done to meet operating goals (Kirwan and Ainsworth 1992). Hierarchical Task Analysis documents system requirements and the order in which tasks must take place.
- Cognitive Task Analysis is a method that focuses on describing the cognitive skills and abilities needed to perform a task proficiently, rather than the specific physical activities carried out in performing the task (e.g., Militello and Hutton 1998). Cognitive Task Analysis is used to analyze and understand task performance in complex real-world situations, especially those involving change, uncertainty, and time pressure.

Example

Two Task Analysis examples are presented for DURESS II: Sequential Flow and Timeline. Figure 10.10 shows a partial normative Sequential Flow Task Analysis for starting up the DURESS II process (Hajdukiewicz and Vicente in press). This task requires the actor to take the process from a state where all of the components are off and the reservoirs are empty to a steady state where the demand and temperature goals for both reservoirs are all satisfied. If the goal to be achieved was a different one (e.g., shut down the process), then the results of the Task Analysis could be different.

Starting from the top left of Figure 10.10, the operator will need to first check the state of the work domain. If the target water demand exceeds the combined capacity of the two feedwater streams, then the task cannot be performed. Next, the states of the valves, pumps, and reservoirs are checked and adjusted to their appropriate settings for startup. Next, the operator must check the level in the reservoir until it exceeds a minimum of its capacity. At that point, the operator checks the demands again and sets the output valves to their respective demand values. Once set, the input valves and pumps need to be configured based on the specific demands. Three categories of configurations are possible depending on the values of the demands (D1 + D2 < 10; D1 + D2 > 10, D1 and D2 < 10; D1 or D2 > 10). After making the appropriate input valve and pump adjustments, the levels in the reservoirs are checked to ensure that they are within a safe range. If they are, then the startup task is complete.

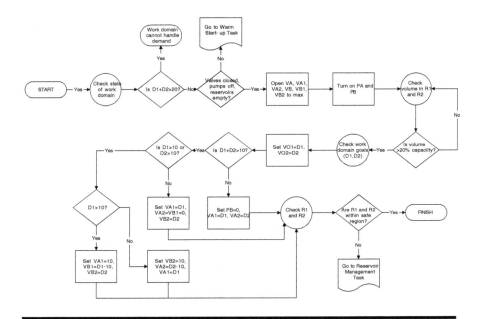

Figure 10.10 An example of a Sequential Flow Task Analysis for starting up the process for DURESS II. (From Hajdukiewicz, J. R. and Vicente, K. J. (forthcoming) A theoretical note on the relationship between work domain analysis and task analysis, *Theoretical Issues in Ergonomics Science.* **Reprinted with permission.)**

Figure 10.11 shows a partial Timeline Task Analysis. The vertical axis includes all components that contribute for water inflow. The horizontal axis represents time. The shaded bars present the component settings that are not zero or off. For example, the first component to be used is VA1 with the setting of 10. After a period of time, the setting for VA1 is changed to 7.

Discussion

Hajdukiewicz and Vicente (in press) discussed the connections between WDA (used in EID) and Task Analysis. They argued that the relationship involves five discrete transformation steps from WDA to Task Analysis:

1. Complete work domain structure
2. Relevant work domain structure
3. Utilized work domain structure
4. Current and desired work domain state
5. Final set of actions and work domain states (outputs of a Task Analysis)

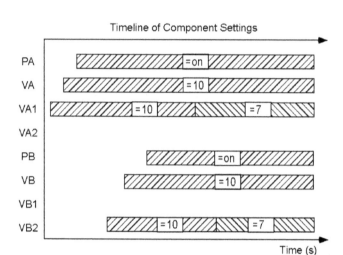

Figure 10.11   An example of a Timeline Task Analysis for DURESS II.

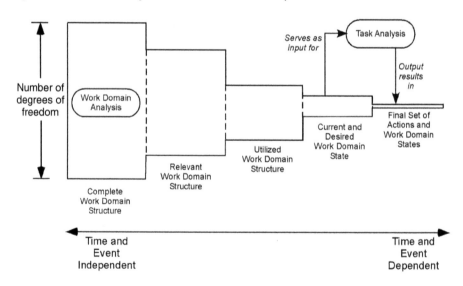

Figure 10.12   Relationship between WDA and Task Analysis. (From Hajduk-iewicz, J. R. and Vicente, K. J. (forthcoming) A theoretical note on the relationship between work domain analysis and task analysis, *Theoretical Issues in Ergonomics Science*. Reprinted with permission.)

In Figure 10.12, each level is connected by a discrete transformation. As transformations move from left to right, the number of degrees of freedom in controlling the system decreases, and time and context dependence

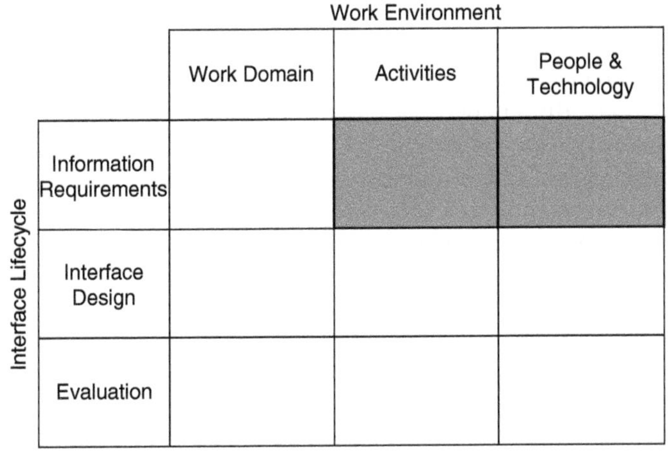

Work Environment

	Work Domain	Activities	People & Technology
Information Requirements			
Interface Design			
Evaluation			

(Interface Lifecycle)

Figure 10.13 Mapping of Task Analysis to the design framework.

increases. In the final transformation, Task Analysis is a function that maps current states onto desired states via a set of human or automated actions.

Mapping onto the Framework

As shown in Figure 10.13, Task Analysis fits onto the design framework in the Information Requirements row. Its focus is mainly to assess the activities, and the people and technology used to perform these activities. In doing so, constraints of the environment may be indirectly considered, but they are not usually expressed and formally modeled.

Situation Awareness Analysis

Situation Awareness has become a significant design goal for a number of industries (Endsley and Garland 2000). Situation Awareness is the experience of fully understanding what is going on in a given situation, seeing each element within the context of the overall goal, and having all the pieces fit together into a coherent picture. It is defined in terms of goals and decision tasks for the particular job. Endsley (2000) discusses the criteria for achieving Situation Awareness and a number of analysis and measurement methods. The nature and mechanisms for achieving Situation Awareness can be described generically as three levels:

- *Level 1 Situation Awareness: Perception.* Perception of and attending to cues is critical for Situation Awareness to be achieved. Misperception or other errors in perception can significantly decrease the chances of Situation Awareness.
- *Level 2 Situation Awareness: Comprehension.* Comprehension relates to the integration of multiple pieces of information and determination of the relevance to the person's goals.
- *Level 3 Situation Awareness: Projection.* Projection relates to the ability to forecast future events and dynamics and relate these to the current state of the process and activities.

Time is a critical aspect of understanding Situation Awareness. Users need to know the time-based dynamics of the situation, including how much time is required to perform tasks and how much time is left before an event occurs.

Current methods for assessing Situation Awareness include the Situation Awareness Global Assessment Technique (SAGAT; Endsley 1988); Situational Awareness Rating Technique (SART; Selcon and Taylor 1990); Situation Awareness–Subjective Workload Dominance (SA-SWORD; Vidulich and Hughes 1991); Crew Situational Awareness (Moiser and Chidester 1991); and Situational Awareness Rating Scales (SARS; Waag and Houck 1994).

Example

DURESS II can be studied heuristically on the requirements that are needed to achieve situation awareness:

- *Level 1 Situation Awareness: Perception.* DURESS II has constraints and relations that need to be displayed consistently to provide the user with cues for action. The EID interface has some perceptual elements, with consistent relations providing these cues for action. For example, the cue for steady state mass in a reservoir is a straight line connecting mass inflow to mass outflow.
- *Level 2 Situation Awareness: Comprehension.* With DURESS II, the constraints and relations link to laws and principles associated with the functioning of the simulator. These then constrain or enable a person to reach the system goals. Comprehension of how this information is relevant to system goals needs to be conveyed in the interface, in the decision support tool, or through training. With the EID interface, the constraints, relations, and principles are shown directly on the display. Users are able to directly see the integration of multiple pieces of information and determine the relevance to goals.

■ *Level 3 Situation Awareness: Projection.* With DURESS II, the ability to forecast future events and dynamics in relation to the current state of the process can be implemented with the current state and derivative information, or with the use of predictive and forecasting methods. In the EID interface, projection of future state from current state is shown in a limited way by the perception of state changes over time and instantaneous derivative information (e.g., mass and energy relationships). For example, operators may predict the rate at which the tank will fill in the future based on this information; however, prediction of unanticipated events cannot be determined beforehand.

Discussion

The Situation Awareness method provides a complementary approach to EID. It focuses on the criteria for interfaces, team environments, and training in improving the awareness of people and teams of the work environment conditions. Situation Awareness provides principles and insights for what needs to be displayed and how it can be displayed for greater awareness.

Mapping onto the Framework

As shown in Figure 10.14, Situation Awareness generally maps onto the Information Requirements and Evaluation rows of the framework. These tools provide insight into the requirements for interface design indirectly.

Contextual Inquiry and Design

Contextual Inquiry and Design are a collection of tools that help the analyst get a better understanding of the users and the reality of their work environments for the purposes of design (Beyer and Holtzblatt 1998). There are four principles to Contextual Inquiry:

■ *Context:* This principle prescribes that analysis is to be performed as close to the actual environment as possible.
■ *Partnership:* This principle involves a collaboration between the analyst and user.
■ *Interpretation:* This principle involves integrating the collected data and what it implies about work structure and possible supporting systems.
■ *Focus:* This principle defines the point of view an interviewer takes when studying the work.

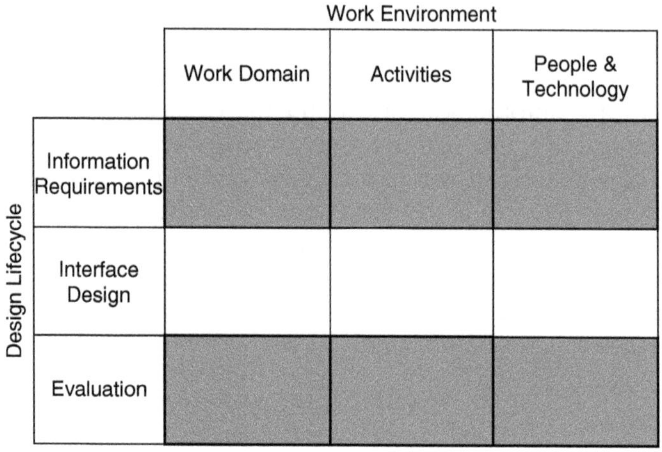

Work Environment

	Work Domain	Activities	People & Technology
Information Requirements			
Interface Design			
Evaluation			

Design Lifecycle

Figure 10.14 Mapping of Situation Awareness on the comparison framework.

Typically five work models are created with Contextual Inquiry (Beyer and Holtzblatt 1998). Each describes a different description of the work environment:

- *Flow Model:* This model captures how work is divided up and coordinated across people to ensure the whole job gets done.
- *Sequence Model:* This model captures how tasks are ordered.
- *Artifact Model:* This model captures the things people create, use, and modify in the course of doing work. This model includes the information presented by the object, parts of the object, the structure of the object as given or based on usage, annotations, presentation of the object, conceptual distinctions, usage of the object, and breakdowns associated with the object.
- *Cultural Model:* This model defines the expectations, desires, policies, values, and approach people take in their work.
- *Physical Model:* This model describes the physical environment in terms of enabling work and constraining work.

Example

When applying Contextual Inquiry and Design to DURESS II, we make the assumption that DURESS II is a real environment operated by a team within a plant as described earlier in this chapter.

- *Flow Model:* If we assume DURESS II is operated by two users in a control room organization, the flow model may look like Figure

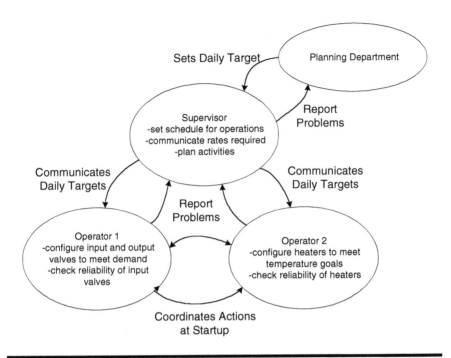

Figure 10.15 Example of flow model for DURESS II with two operators running the simulation, a supervisor overseeing operations and a planning department setting targets.

10.15. Operators 1 and 2 coordinate actions during startup. The supervisor sets the schedule for operations, communicates daily targets, and receives information regarding the status and any problems associated with startup. In doing so, the supervisor interacts with the planning department.

- *Sequence Model:* With DURESS II, the sequence model shows the steps taken to get to a goal state from an initial state. Figure 10.16 shows an example model. Each phase in the sequence has an intent or goal. The sequence is to manipulate the equipment variables to achieve that intent and ultimately reach the goal state.
- *Artifact Model:* With DURESS II, this model captures the things people create, use, and modify in the course of doing work. This can include both how users interact with the interface forms available and the tools users create to operate the simulation. Figure 10.17 shows two examples within the artifact model, one that is directly on the EID display and the other that is available as a heuristic tool but not on the display. The oval in Figure 10.17 highlights the emergent feature line that connects mass inflow to outflow for Reservoir 1. When the line is straight, mass inflow is balanced with outflow. Users can get

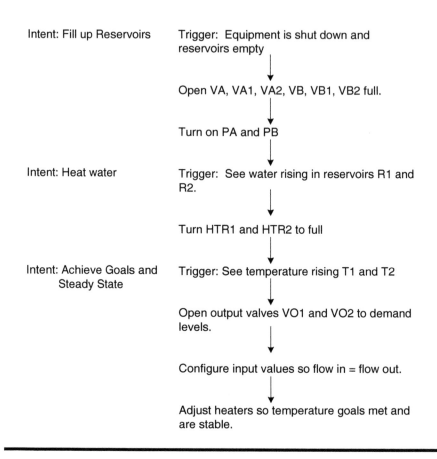

Intent: Fill up Reservoirs

Trigger: Equipment is shut down and reservoirs empty

Open VA, VA1, VA2, VB, VB1, VB2 full.

Turn on PA and PB

Intent: Heat water

Trigger: See water rising in reservoirs R1 and R2.

Turn HTR1 and HTR2 to full

Intent: Achieve Goals and Steady State

Trigger: See temperature rising T1 and T2

Open output valves VO1 and VO2 to demand levels.

Configure input values so flow in = flow out.

Adjust heaters so temperature goals met and are stable.

Figure 10.16 Sequence model applied to DURESS II.

attuned to this artifact perceptually and control the simulation based on the graphic feature. The box on the left side shows a side artifact that is not part of the EID display, a set of equations that result in a steady-state mass balance under normal conditions.

■ *Cultural Model:* Figure 10.18 shows an example of a cultural model applied to DURESS II under the assumptions previously identified. Within plant operations, there can be a safety and efficiency focus driven by the operations culture that permeates through different plant roles: supervisor and operators. This model defines the expectations, desires, policies, values, and approach people take in their work.

■ *Physical Model:* Figure 10.19 shows the physical environment of the DURESS II environment. It shows the physical interactions and constraints in the work environment.

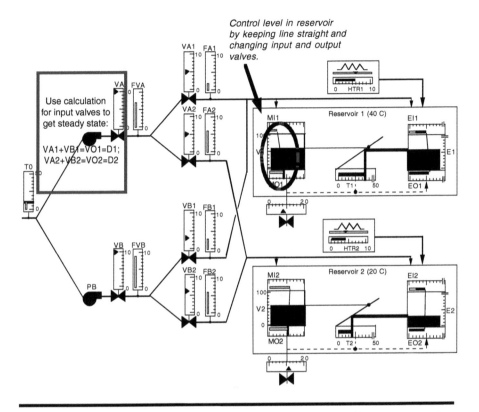

Figure 10.17 Examples of calculation and interface form artifacts to control levels in DURESS II.

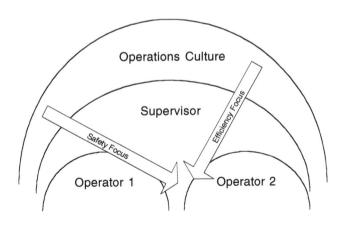

Figure 10.18 An example of the cultural model applied to DURESS II.

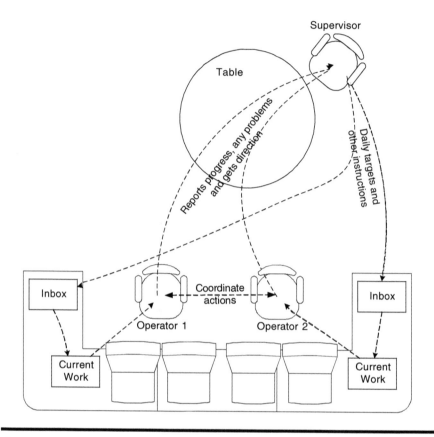

Figure 10.19 Physical model associated with the hypothetical control room environment.

Discussion

Contextual Inquiry provides great insight into the work context associated with socio-technical structures and activities. It provides a way to analyze the immediate work environment of the users and the influences of organizations and culture. The EID approach does not cover the socio-technical structure in the same level of detail; it focuses on the work domain that needs to be controlled. Thus, these approaches can be complementary in developing new insights for design.

Mapping onto the Framework

As shown in Figure 10.20, contextual inquiry analysis generally maps onto the Information Requirements row of the framework for all elements of the work environment. The Physical Model maps onto the work domain

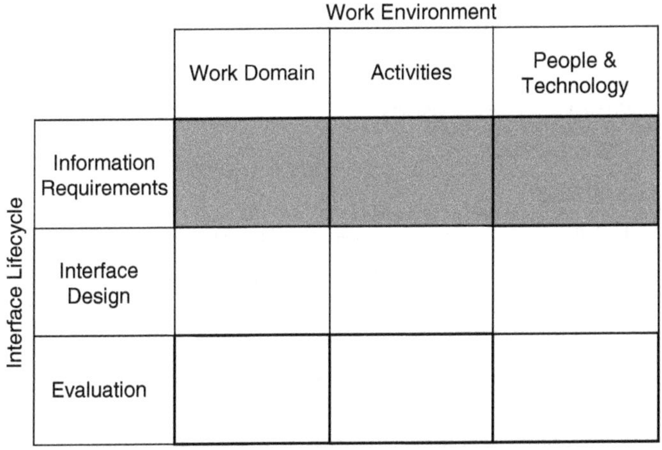

Work Environment

	Work Domain	Activities	People & Technology
Information Requirements			
Interface Design			
Evaluation			

(left vertical axis label: Interface Lifecycle)

Figure 10.20   The mapping of Contextual Inquiry onto the design framework.

element. The Sequence Model maps onto the Activities element. The Cultural Model, Flow Model, and Artifact Model map onto the People and Technology element.

GOMS

GOMS is an acronym that stands for Goal, Operators, Methods, and Selection Rules (Card et al. 1983). A GOMS model is composed of methods that are used to achieve specific goal states that the user wants to occur. The GOMS approach results in a sequence of operators at the lowest level (perceptual, cognitive, and motor) that can bring about changes to the work environment. The operators are specific steps that a user performs and are assigned a specific execution time. If a goal can be achieved by more than one method, then selection rules are used to determine the proper method.

Example

Figure 10.21 is an example of the implementation of GOMS with the DURESS II for startup.

Discussion

GOMS is very effective at modeling predictable activities and the efficiencies in carrying out these activities. However, there have been a number of noted limitations (Card et al. 1980). First, the model is applied to skilled

```
STARTUP

GOAL: Startup DURESS II safely
METHOD: Open input valves to fill tanks
OPERATOR: VB1=5, VB2=5, VB=10, turn on PB
OPERATOR: VA1=5, VA2=5, VA=10, turn on PA
METHOD: Turn on heaters to heat water
OPERATOR: HTR1=10, HTR2=7


STABILIZATION OF FLOW
GOAL: Stabilize flow and meet demand for outflow VO1=D1, VO2=D2
SELECTION RULE: IF D1+D2>20, shut down system, ELSE configure input valves
METHOD (Shut down system): Turn off heater and pumps
OPERATOR: PA, PB off, VA=VA1=VA2=VB=VB1=VB2=0, HTR1=HTR2=0
METHOD (Configure system): Look at VO1's goal and adjust VA1+VB1=VO1, try to even out VA1and
VB1 (i.e., 50% each). Look at VO2's goal and adjust VA2+VB2=VO2, try to even out VA2and VB2 (i.e.,
50% each).
OPERATOR: VA1+VB1=VO1, VA1=VB1, VA2+VB2=VO2, VA2=VB2,
OPERATOR: When R1/R2 reaches about 20 units, turn VO1 and VO2 to goal flow.


STABILIZATION OF HEAT
GOAL: Stabilize temperature to goal settings (T1=T1g, T2=T2g)
METHOD: Calculate steady-state temperature
OPERATOR: VO1X(T1-To)=VO1X30; VO2X(T2-To)=VO2X10 – of total heat required divide by 30.
i.e., HTR1=VO1X30/30=VO1, HTR2=VO2X10/30=VO2/3 is the final steady state setting of the heaters
```

Figure 10.21   GOMS example for the DURESS II startup.

users, not to beginners or intermediates. Second, the model doesn't account for either learning of the system or its recall after a period of disuse. Third, the model doesn't address functionality of the work domain.

In contrast, EID supports both novice and expert behavior and allows for learning and problem solving. In addition, EID displays the functional structure of the work domain.

Mapping onto the Framework

As shown in Figure 10.22, GOMS generally maps onto the Information Requirements — Activities cell of the framework. It can be very specific in terms of activities, their sequences, and task efficiencies.

Use Case Scenario Design

Use Cases provide a perspective to system design and development based on user-interaction scenarios (Carroll 1995). They provide a description of an activity that the user is engaged in when performing a task; this description is detailed enough to eventually make decisions regarding design considerations. These scenarios can be narrative descriptions of what people do and experience as they try to make use of computer systems and applications. They tend to capture contextual information, issues, and salient events. Scenarios can also identify individual motivations and expectations towards the system, characterize organizational influences

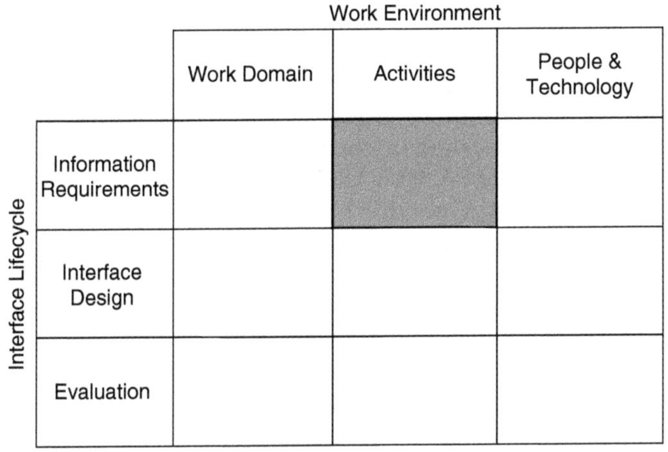

Work Environment

	Work Domain	Activities	People & Technology
Information Requirements			
Interface Design			
Evaluation			

Interface Lifecycle

Figure 10.22   Mapping GOMS to the design framework.

and culture, identify actions taken with associated reasoning, and characterize results.

Use Cases guide the design of the user interface and the underlying architecture that supports the user interface, as well as the usability testing of the system throughout the design process. In addition, users participate at various stages of the development process to provide feedback on the product requirements and design to ensure that the design fits their mental model (i.e., their expectations of how the product should operate) and meets their task needs.

Use Case Scenarios can take various forms, including textual narratives, storyboards of annotated panels, video mockups, scripted prototypes, or physical situations that support certain user activities.

Example

With DURESS II, an example of part of a Use Case Scenario for startup using the Unified Modeling Language (Object Management Group 2003) is shown in Table 10.1.

Discussion

Use Case Scenarios provide a basis for system design by describing the activities and context around work situations. The Use Case does not explicitly model the functional structure of the work domain, but analyzes activities as well as the roles of people and technology in meeting the required goals.

Table 10.1 Example of Part of a Use Case Scenario for DURESS II Startup

Use Case	Startup (1)
Description	The goal of this use case is to fill water in the reservoirs and start heating up the water Sources: PA, VA, VA1, VA2, PB, VB, VB1, VB2, R1, R2, VO1, VO2
Actors	Primary: Operator 1, Operator 2 Secondary: Supervisor
Assumptions	Reservoir is filling with water and water temperature is rising
Steps	<1> Operator 1 — Open all input valves fully (VA = VB = VA1 = VA2 = VB1 = VB2 = 10) <2> Operator 1 — Turn on pumps (PA = PB = on) <3> Operator 2 — See water in reservoirs (R1 and R2) at 10% and filling <4> Operator 2 — Turn on heaters (HTR1 = HTR2 = 10)
Variations	<1> Operator 1 — Open all input valves halfway (VA=VB=VA1=VA2=VB1=VB2=5) <4> Operator 2 — Turn on heater settings to halfway (HTR1 = HTR2 = 5)
Nonfunctional	<Performance>: <Goal state achieved within 30 minutes> <Priority>: <Ensure D1, T1 are met as *a priority*>

Mapping onto the Framework

As shown in Figure 10.23, Use Case Scenarios fit onto the design framework primarily in the Information Requirements and Evaluation rows. Their focus is mainly to assess the activities of users and the applications' ability to support these activities. In doing so, implicit constraints of the work environment are modeled. In addition, allocation of people and technology is generally performed for the activities.

Participatory Design

Participatory Design is an approach for the assessment, design, and development of technological and organizational systems that places a premium on the active involvement of workplace practitioners (usually potential or current users of the system) in design and decision-making processes (Muller et al. 1992). Two techniques have been popular in Participatory Design: CARD and PICTIVE.

CARD combines some elements of storyboarding with dynamic, participatory attributes of games for design (Muller et al. 1995). Participants describe and critique the task at a high level by manipulating cards and creating new ones. With this approach cards can be developed to represent

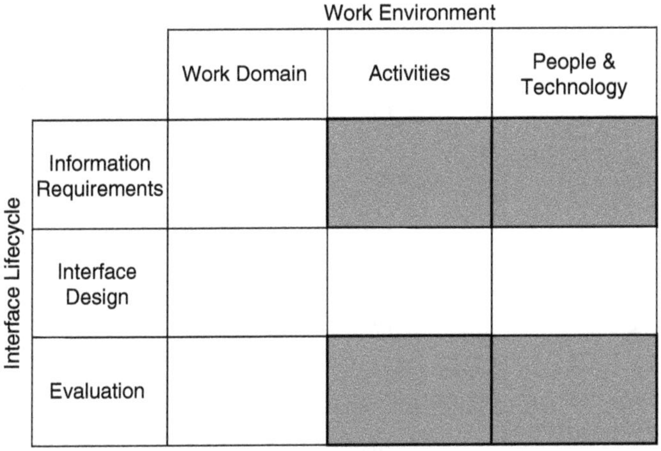

Work Environment

	Work Domain	Activities	People & Technology
Information Requirements		▓	▓
Interface Design			
Evaluation		▓	▓

(rows grouped under "Interface Lifecycle")

Figure 10.23 Mapping Use Case Scenario Design onto the design framework.

the assumed cognitive or motivational state of the user. Participants can agree on the user goals for certain work scenarios and can position where in the sequence the goal is determined. User strategies can also be captured for achieving the goal at relative positions in the sequence. The user's changing understanding of the situation can be represented as needed.

PICTIVE is a technique that uses known objects to represent components of computer systems to the level of detail of display design (Muller et al. 1995). Participants manipulate these objects to describe their ideas about the design, and the work context behind the design.

Example

Figure 10.24 shows an example of a display that could have resulted from a PICTIVE exercise. Users would have knowledge of the physical structure of the DURESS II system and interconnection of components. The non-EID interface could have resulted from users creating the display based on their knowledge of equipment in the plant.

Discussion

Participatory methods provide an efficient and democratic set of tools for determining information requirements, interface design, and evaluation elements associated with activities. Through the process, work domain, people, and technology aspects can be captured indirectly. In contrast, EID is more focused at information requirements and modeling at the work domain element to create displays.

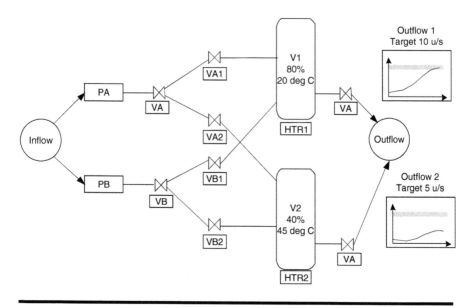

Figure 10.24 Example of a display that could have been the output of an initial PICTIVE exercise.

Mapping onto the Framework

As shown in Figure 10.25, Participatory Design covers primarily the Activities element of the work environment across all elements in the interface lifecycle. Many of the results are not detailed models, but can be created very efficiently.

Applying User Interface Design Principles

A number of principles have been developed to aid designers in their quest to create effective interfaces. This section describes a few popular principles. This is only a subset of many principles that have been identified. For a more thorough treatment of this topic, refer to Galitz (1997) for some examples.

Once information requirements have been identified, good user interface design transforms these requirements to display elements and organizes them in a way that is compatible to how users perceive and utilize information based on the work context and conditions of use. The following items are some of the more familiar principles used in interface design (e.g., Galitz 1997; Horn 1998):

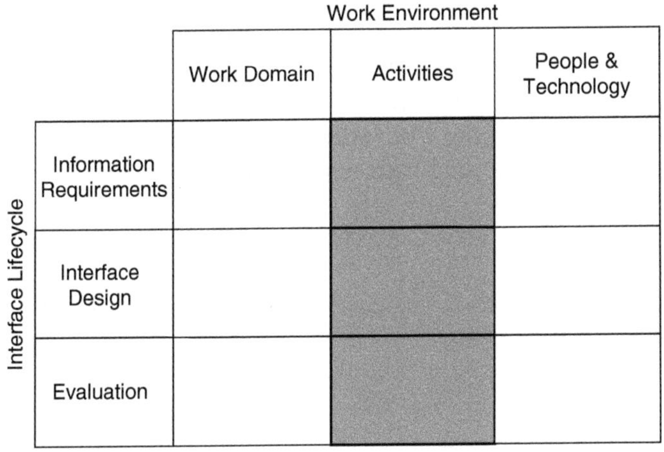

Work Environment

	Work Domain	Activities	People & Technology
Information Requirements			
Interface Design			
Evaluation			

Interface Lifecycle

Figure 10.25 Mapping Participatory Design onto the design framework.

- *Consistency:* The interface design should be consistent with the experiences conforming to work conventions. This will minimize use errors in navigation, information retrieval, and action execution.
- *Upper-left starting point:* Based on the western cultural convention, people tend to first start looking at a display in the upper-left corner. Workflow and causal relationships should be mapped to screen layout and conform to this convention.
- *Navigation:* In addition to being consistent, navigation within and across screens should match workflow and expectations.
- *Visually pleasing composition:* A number of principles have been identified as important for creating visually pleasing interfaces. These include balance, symmetry, regularity, predictability, sequentiality, economy, unity, proportion, simplicity, and groupings. When deciding on the organization of visual elements, these need to be taken into account.
- *Grouping:* Gestalt principles provide six general ways to group screen elements spatially: principle of proximity, principle of similarity, principle of common region or closed form, principle of connectedness, principle of good continuation, and principle of closure.
- *Amount of information:* This should be calibrated to that which is required for the task at hand. Too little information could result in important information that is hidden from the user and awkward to access; too much information could result in confusion. The amount of information that is displayed could be increased by chunking data

on the display. Using Miller's principle that people cannot retain more that 7 ± 2 items in short-term memory, chunking data using this heuristic could provide the ability to add more relevant information on the screen as needed.

■ *Meaningful ordering:* The ordering of elements and their organization should have meaning with respect to the task and information processing activities.

■ *Distinctiveness:* Screen objects should be distinguishable from other objects.

■ *Focus and emphasis:* Salience of objects should reflect the relative importance of focus.

Example

The principles can be applied to the EID interface of DURESS II in terms of screen organization. The following list describes how a few of these principles were applied:

■ *Consistency:* Objects such as input valves have a consistent appearance and behavior. The arrow on the valve object controls the valve setting.

■ *Upper-left starting point:* The EID interface is organized causally from left to right. The first objects are the input valves, then come the reservoirs, then finally the output valves.

■ *Navigation:* The EID interface only has one screen. Navigation is performed by visually scanning among objects on the display. The user can follow the causal network through the visual interconnection of elements. For example, pump PA generates flow through VA, VA1, VA2. In addition, perceptual navigation can result across functional levels of abstraction to system purposes. For example, heater settings link to heat flow that is adjacent to them; heat flow is an input to energy input that has a relation with temperature.

■ *Visually pleasing composition and amount of information:* With the EID interface there is balance and symmetry across screen elements. A trade-off was made in design. The additional functional information made the display more complex; however, this information provides valuable cues for detection, diagnosis, and action planning.

■ *Grouping:* A couple of examples of how the Gestalt principles were applied to the EID interface are with the principles of proximity and connectedness. For proximity, the input valve setting graphic is adjacent to the input valve flow graphic. For connectedness, the graphic lines show the causal relationships among input valves.

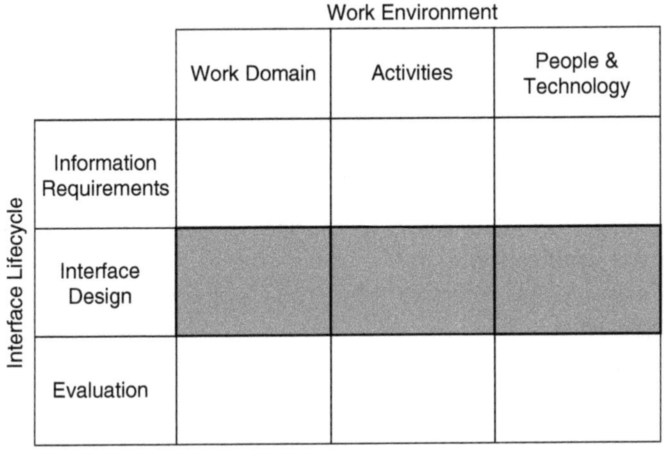

Work Environment

	Work Domain	Activities	People & Technology
Information Requirements			
Interface Design	▓▓▓	▓▓▓	▓▓▓
Evaluation			

Interface Lifecycle

Figure 10.26 Mapping the use of User Interface Principles onto the design framework.

- *Focus and emphasis:* The most distinctive and salient feature of the EID interface is the emergent feature graphics on the right-hand side. The size and shapes of the emergent feature graphics form the attention of the user to this area. From experiments it was determined that these emergent graphics were useful in helping people cope with unanticipated situations.

Discussion

Once information requirements have been generated, User Interface Design Principles provide many great suggestions on how to create, organize, and display graphic forms that will improve the final design. EID helps to initially generate and structure these requirements, and define the constraints on graphic forms. User Interface Design Principles help bridge the remaining gaps.

Mapping onto the Framework

As shown in Figure 10.26, the use of User Interface Principles generally map at each work environment element in the Interface Design stage of the lifecycle.

Usability Evaluation

Usability Evaluation is a popular approach predominantly used for the evaluation of designs (Nielsen 1993). Usability Evaluation is a user-centered, user-driven method that defines the system as the connection between the human user and the application software and hardware. It focuses evaluation on whether the system delivers the right information in the appropriate way for users to complete their tasks.

Typically there are nine usability areas considered when evaluating interface designs with users (Nielsen 1993):

- *Terminology:* This refers to the labels, acronyms, and terms used in the application.
- *Workflow:* This refers to the natural sequence of tasks in using the application.
- *Navigation:* This refers to the methods used to navigate to various parts of the application.
- *Symbols:* This refers to symbols and icons used in the application to convey information and status.
- *Access:* This refers to the availability and ease of access of information and functions to the user.
- *Content:* This refers to the content of information available to the user.
- *Format:* This refers to the format in which the content is conveyed to the user.
- *Functionality:* This refers to the specific functionality of the application and its usefulness to the user.
- *Organization:* This refers to the layout of the application screens.

The usability criteria typically used for evaluation are noted below for the areas above:

- *Visual clarity:* This dimension is concerned with the way in which information is displayed on the screen or paper document. Information should be clear, well-organized, unambiguous, and easy to read. This will enable users to find required information, draw the user's attention to important information, and allow the user to see where information should be entered on the screen, quickly and easily.
- *Consistency:* This dimension conveys that the way the system looks and works should be consistent at all times. Consistency reinforces user expectations by maintaining predictability across the interface.
- *Compatibility:* This dimension corresponds to the extent the interface conforms with existing user conventions and expectations. If the interface components are familiar to the users, it will be easier for them to navigate, understand, and interpret what they are looking at and what the system is doing. Examples include the use of menus with "File, Edit, View …."
- *Informative feedback:* Users should be given clear, informative feedback on where they are in the system, what actions they have taken,

whether the actions have been successful, and what actions should be taken next.

■ *Explicitness:* The way the system works and is structured should be clear to the user.

■ *Appropriate functionality:* The system should meet the requirements and needs of the users when carrying out tasks.

■ *Flexibility and control:* The interface should be sufficiently flexible in structure, in the way information is presented and in terms of what the user can do, to suit the needs and requirements of all users, and to allow them to feel in control of the system.

■ *Error prevention and correction:* The system should be designed to minimize the possibility of user error, with built-in functions to detect when these errors occur. Users should be able to check their inputs and to correct errors and potential error situations before the input is processed.

■ *User guidance and support:* Informative, easy-to-use, and relevant guidance should be provided, both on the computer and hardcopy versions, to help the users understand and use the system.

■ *System usability problems:* Problems associated with using the system should be minimized.

Examples

When conducting a usability assessment of DURESS II, the following are sample questions that could be asked with sample responses based on the EID interface in Figure 10.5:

Visual Clarity

■ Is information organized logically on the screen? The layout of the EID interface shows the causal structure of the microworld simulation environment. Functional information is collocated with relevant components.

■ Are different types of information clearly separated or differentiated? The EID interface shows different types of information by using different forms (e.g., setting bar for valves, indicator for flow).

■ Does the screen document appear cluttered? The EID display seems cluttered compared with the non-EID display. Salience of critical information could be achieved by manipulating color intensity.

Consistency

- Are icons, symbols, graphical representations, and other pictorial information used consistently? In the EID interfaces, graphical items are used consistently. For example, flow rates have a bar indicating their state with respect to a scale, and mass/energy relationships are shown consistently.

- Is the method for entering information consistent throughout? To make control actions using the EID interface, the user must select a control triangle and drag it to the desired end-point. This is consistent for all components in DURESS II.

- Is the action required to move the cursor around the display consistent? It is consistent with the use of a mouse.

Compatibility

- When a user makes an input movement in a particular direction, does it map directly to the direction of movement on the screen? Yes, with the use of a mouse.

- Is the graphical display compatible with the users' view of what they are representing? With the EID interface, the graphical display shows more functional information.

Informative Feedback

- Are changes that occur as a result of user action/input clearly displayed? The EID interface shows the dynamics associated with control action changes.

Explicitness

- Is it clear what stage the system has reached in a task? The EID interface provides feedback on when the system has reached steady state.

- Is it clear how, where, and why changes in one part of the system affect other parts of the system? The interface shows the causal links between different parts of the system, allowing the user to determine how changes in one part of the system affect other parts of the system.

Appropriate Functionality

- Is the input device available to the user appropriate for the tasks carried out? The mouse provides an efficient method to make changes to the components.

- Does each display contain all the information that the user feels is relevant to the task? The EID interface contains the functional information linking component actions to higher level purposes. It does not contain task-specific information on the sequencing of actions.

- Does the system allow users to do what they feel is necessary in order to carry out the task? The DURESS II system has the flexibility to allow users to carry out the task of startup in many ways. If problems occur, alternative ways of controlling the system may be used to some extent to reach the goals.

Flexibility and Control

- Are shortcuts available as appropriate? Shortcuts, such as auto-configuration, are not available to the user.

- When appropriate, do users have control of the order in which they request information, or carry out a series of activities? Yes, with DURESS II system, users are able to control the order in which they carry out a series of activities.

- Is the user able to tailor aspects of the interface for their own preferences or needs? With the EID interface, users cannot tailor the interface to their own preferences.

Error Prevention and Correction

- Does the system ensure the user double-checks any requested actions that may be catastrophic if requested unintentionally? No, users can perform catastrophic actions without being warned with DURESS II.

- When system errors occur, does the user get access to diagnostic information to resolve the problem? When there is a malfunction with DURESS II, users need to resolve this perceptually with the interface forms.

User Guidance and Support

- Is help available on the system? No, only hardcopy guides are available.

- Do hardcopy guides (user manuals) provide an in-depth, comprehensive description, covering all aspects of the system? Yes, the DURESS II system has system documentation and an explanation of how the interface works.

- Is the organization of all forms of the user guide related to the tasks the user needs to carry out? No, the user guide describes the functioning of DURESS II and the interfaces, but not the tasks to be performed explicitly.

System Usability Problems

- Is guidance on using the system clear and accessible to the user? DURESS II has a training module for using the system that is clear and accessible.

- Is the help function flexible? There is no help function with DURESS II.

- Does the system require users to remember too much information while carrying out a task? No, the functional information relating component actions to goals is visible to the user, minimizing the need for remembering these relationships.

Discussion

Usability Evaluation is primarily focused on evaluating interface designs once they have been developed. The approach does not generally capture information requirements, but can feed back the appropriateness of these requirements after evaluation with users. As such, Usability Evaluation is complementary to EID. EID requires an evaluative aspect as part of the design process lifecycle to create appropriate designs.

Mapping onto Framework

As shown in Figure 10.27, Usability Evaluation generally maps onto the Evaluation row of the design framework.

SUMMARY

The purpose of this chapter was to provide some insight into using EID with other methods, with the ultimate goal of creating effective interface

Work Environment

	Work Domain	Activities	People & Technology
Information Requirements			
Interface Design			
Evaluation			

Interface Lifecycle

Figure 10.27 Mapping Usability Evaluation onto the design framework.

Work Environment

		Work Domain	Activities	People & Technology
Interface Lifecycle	Information Requirements	EID, Cognitive Work Analysis, Situation Awareness Analysis, Contextual Inquiry	Cognitive Work Analysis, Task Analysis, Situation Awareness Analysis, Contextual Inquiry, GOMS, Scenario-based Design, Participatory Design (CARD)	EID, Cognitive Work Analysis, Situation Awareness Analysis, Contextual Inquiry, Scenario-based Design
	Interface Design	EID, User Interface Design Principles	Participatory Design (PICTIVE), User Interface Design Principles	EID, User Interface Design Principles
	Evaluation	Situation Awareness Analysis, Usability Evaluation	Situation Awareness Analysis, Scenario-based Design, Participatory Design, Usability Evaluation	Situation Awareness Analysis, Scenario-based Design, Usability Evaluation

Figure 10.28 Mapping all the chosen methods onto the design framework.

designs. We introduced a framework (Figure 10.3) that integrated work environment elements (work domain, activities, people and technology) with the design lifecycle elements (information requirements generation, interface design, and evaluation). Figure 10.28 maps all the discussed methods onto the framework.

It is unlikely that EID (or any other method) will efficiently and effectively provide the entire answer to the interface design question in real, complex environments. Other methods are required to fill in the gaps for effective final designs. However, EID does provide alternative techniques and helps fill in some gaps. For example, the Abstraction Hierarchy is an alternative to some of the models in Contextual Inquiry. In addition, EID helps fill in some gaps in creating interfaces for complex work environments.

11

CONCLUSION

In summary, we have tried to convey several key ideas in this book. First, *WDA is an effective way to understand a complex domain better.* There are many useful techniques for completing a WDA, with numerous case studies in this book. The analysis itself should not be considered an output that is used only once. Much can be gained by revisiting the analysis several times to understand different variables, their availability, and how they relate to each other. An analysis may also be a living entity, changing in scope as needs change and providing a basis to assess new technologies. You have never wasted time by learning more about your work domain. We also encourage you to apply WDA to complex domains that are not included in this book. You may find that the case studies featured in the book can give you some ideas on how to handle the analysis for the new domain.

Interface design is an art, but a learnable art. You can use your analysis to systematically identify the variables you need to display, find the key relationships between them, and begin to structure your display. The visual thesaurus provides a lookup table of basic visual elements mapped to variable requirements. For richer examples, take advantage of the case studies in the book. They show actual implementations of many of the elements from the visual thesaurus. Interface design is often a process of borrowing from other designs and adapting them to suit your new situation. Keeping track of many visual examples can provide you with a tool bag of possible design ideas. For new visualizations there are sources as well that you can take advantage of, ranging from understanding the mathematics or patterns in the data, to textbook graphs and models.

EID is an effective addition to other design approaches. It provides an analysis that is useful for determining work domain constraints and provides a visual basis for displaying that information. The most effective designs will take advantage of EID and build on it using other design approaches that fill in the gaps by capturing different aspects of the work environment and interface lifecycle process. In particular, the final measure of your design is how well it supports human performance. This demands human performance testing and usability testing (these topics are not the focus of EID, but are covered in other texts).

Finally, we encourage you to take the ideas presented in the book, try them out on design problems that are new or variations of the featured case studies, generate new design ideas that expand on the visual thesaurus, and continue to learn by exploration.

REFERENCES

Amelink, M. H. J. (2002) "Visual control augmentation by presenting energy management information in the primary flight display: An ecological approach," unpublished thesis, Technical University of Delft. Available at www.amelink.net.

Amelink, M. H. J., van Paassen, M. M., Mulder, M., and Flach, J. M. (2003) "Applying the abstraction hierarchy to the aircraft manual control task," *Proceedings of the 12th Annual Symposium on Aviation Psychology*, April 14–17, 42–47.

Beltracchi, L. (1989) "Energy, mass, model-based displays, and memory recall," *IEEE Transactions on Nuclear Science,* 36:1367–1382.

Beltracchi, L. (1995) "An OLE interface concept for rankine cycle-based heat engines," paper presented at the Topical Meeting of the American Nuclear Society: Computer Display and Support Systems, Philadelphia.

Beyer, H. and Holtzblatt, K. (1998) *Contexual Design: Defining Customer-Centered Systems,* San Francisco: Morgan Kauffmann.

Bisantz, A. M., Burns, C. M., and Roth, E. M. (2002) "Validating methods in cognitive engineering: A comparison of two work domain models," *Proceedings of the 46th Annual Meeting of the Human Factors and Ergonomics Society,* 521–525.

Bisantz, A. M., Roth, E., Brickman, B., Gosbee, L. L., Hettinger, L., and McKinney, J. (2003) "Integrating cognitive analyses into a large-scale system design process," *International Journal of Human–Computer Studies,* 58:177–206.

Blike, G. T., Surgenor, S. D., and Whalen, K. (1999) "A graphical object display aids anesthesiologists performing two diagnostic tasks compared with a standard display," *Journal of Clinical Monitoring and Computing,* 15:37–44.

Blike, G. T., Surgenor, S. D., Whale, P. K., and Jensen, J. (2000) "Specific elements of a new hemodynamics display improves the performance of anesthesiologists," *Journal of Clinical Monitoring and Computing,* 16:485–491.

Burns, C. M. (2000) "Putting it all together: Improving display integration in ecological displays," *Human Factors,* 42:226–241.

Burns, C. M. and Proulx, P. (2002) Solving problems through interface design. *Ergonomics in Design,* 10(4), 10–16.

Burns, C. M. and Vicente, K. J. (1996) "Comparing the functional information content of displays," *Proceedings of the 28th Annual Conference of the Human Factors Association of Canada,* 59–64.

Burns, C. M., Bryant, D., and Chalmers, B. (2000) "A work domain model to support naval command and control," *Proceedings of the 2000 IEEE International Conference on Systems, Man and Cybernetics,* 2228–2233.

291

Burns, C. M., Bryant, D. J., and Chalmers, B. A. (2001) "Scenario mapping with work domain analysis," *Proceedings of the 45th Annual Meeting of the Human Factors and Ergonomics Society,* 424–428.

Burns, C. M., Kuo, J., and Ng, S. (2003) "Ecological Interface Design: A new approach to visualizing network management," *Computer Networks,* 43:369-388.

Burns, C. M., Thompson, L. K., and Rodriguez, A. (2002) "Mental workload and the display of abstraction hierarchy information," *Proceedings of the 46th Annual Meeting of the Human Factors and Ergonomics Society,* 235–239.

Card, S. K., Moran, T. P., and Newell, A. (1980) "Computer text editing: An information-processing analysis of a routine cognitive skill," *Cognitive Psychology,* 12:32–74.

Card, S. K., Moran, T. P., and Newell, A. (1983) *The Psychology of Human–Computer Interaction,* Hillsdale, NJ: Lawrence Erlbaum Associates.

Carroll, J. M. (Ed.) (1995) *Scenario-Based Design: Envisioning Work and Technology in System Development,* New York: John Wiley & Sons, Inc.

Cebrowski, A. K. and Garstka J. J. (1998) "Network-centric warfare: Its origin and future," *Naval Institute Proceedings.*

Chéry Sandra, S. (1999) "Applying ecological interface design and perceptual control theory to the design of the control display unit," unpublished thesis, University of Toronto.

Chéry, S., Vicente, K. J., and Farrell, P. (1999) "Perceptual control theory and ecological interface design: Lessons learned from the CDU," *Proceedings of the 43rd Annual Meeting of the Human Factors and Ergonomics Society,* 389–393.

Coekin, J. A. (1969) "A versatile presentation of parameters for rapid recognition of total state," *Proceedings of the IEE International Symposium on Man–Machine.*

Dinadis, N. (2002) ProSum Ecological Object Library, personal communication.

Dinadis, N. and Vicente, K. J. (1999) "Designing functional visualizations for aircraft systems status displays," *International Journal of Aviation Psychology,* 9:241–269.

Duez, P. (2003) "Testing the generalizability of ecological interface design to computer network monitoring," unpublished thesis, University of Toronto.

Duez, P. and Vicente, K. J. (submitted) "Ecological interface design and computer network management: The effects of network size and fault frequency," *International Journal of Human–Computer Studies.*

Endsley, M. R. (1988) "Design and evaluation for situation awareness enhancement," *Proceedings of the 32nd Annual Meeting of the Human Factors and Ergonomics Society,* 97–101.

Endsley, M. R. (2000) "Theoretical underpinnings of situation awareness: A critical review," in Endsley, M. R., and Garland, D. J. (Eds.), *Situation Awareness Analysis and Measurement,* 3-32, Mahwah, NJ: Lawrence Erlbaum Associates.

Endsley, M. R. and Garland, D. J. (Eds.) (2000) *Situation Awareness Analysis and Measurement,* Mahwah, NJ: Lawrence Erlbaum Associates.

Galitz, W. O. (1997) *The Essential Guide to User Interface Design: An Introduction to GUI Design Principles and Techniques,* New York: John Wiley & Sons, Inc.

Gibson, J. J. (1979/1986), *The Ecological Approach to Visual Perception,* Hillsdale, NJ: Lawrence Erlbaum Associates.

Gibson, J. J. and Crooks, L. E. (1938) "A theoretical field-analysis of automobile driving," *American Journal of Psychology,* 51:453–471.

Hall, G. W. and Trout, G. M. (1968), *Milk Pasteurization,* Westport, CT: AVI.

Hajdukiewicz, J. R. (2001) "Adapting to change in complex worlds: A study of higher-level control and key success factors in a process control microworld," unpublished thesis, University of Toronto.

Hajdukiewicz, J., Doyle, D. J., Milgram, P., Vicente, K. J., and Burns, C. M. (1998) "A work domain analysis of patient monitoring in the operating room," *Proceedings of the 44th Annual Meeting of the Human Factors and Ergonomics Society,* 1034–1042.

Hajdukiewicz, J. R. and Vicente, K. J. (forthcoming) "A theoretical note on the relationship between work domain analysis and task analysis," *Theoretical Issues in Ergonomics Science.*

Hajdukiewicz, J. R., Vicente, K. J., Doyle, D. J., Milgram, P., and Burns, C. M. (2001) "Modeling a medical environment: An ontology for integrated medical informatics design," *International Journal of Medical Informatics,* 62:79–99.

Hansen, J. P. (1995) "Representation of system invariants by optical invariants in configural displays for process control," in Hancock, P., Flach, J., Caird, J., and Vicente, K. J. (Eds.), *Local Applications of the Ecological Approach to Human–Machine Systems,* 208–233, Hillsdale, NJ: Erlbaum Associates.

Held, G. (1998) *Internetworking LANs and WANs: Concepts, Techniques and Methods, 2nd Ed.,* Chichester, England: John Wiley & Sons.

Horn, R. E. (1998) *Visual Language: Global Communications for the 21st Century,* Bainbridge Island, WA: MacroVU, Inc.

Hunter, C. N., Vicente, K. J., and Tanabe, F. (1996) "Can 'theoretical' training improve fault management performance?," *Proceedings of the 1996 American Nuclear Society International Topical Meeting on Nuclear Plant Instrumentation, Control and Human–Machine Interface Technologies,* 683–690.

Institute of Medicine, Committee on Quality of Health Care in America (2000). *To Err is Human: Building a Safer Health System.* Kohn, L.T., Corrigan, J.M., and Donaldson, M.S., Eds. Washington, D.C: National Academy Press.

Itoh, J., Sakuma, A., and Monta, K. (1995) "An ecological interface for supervisory control of BWR nuclear power plants," *Control Engineering Practice,* 3:231–239.

Jamieson, G. A. (2002) "Empirical evaluation of an industrial application of ecological interface design," *Proceedings of the 46th Annual Meeting of the Human Factors and Ergonomics Society,* 536–540.

Jamieson, G. A. (2003a) "Comparison of information requirements from task- and system-based work analysis," *Proceedings of the International Ergonomics Association XVth Triennial Conference,* Seoul, Korea, unnumbered pages.

Jamieson, G. A. (2003b) "Bridging the gap between cognitive work analysis and ecological interface design," *Proceedings of the 47th Annual Meeting of the Human Factors and Ergonomics Society,* 273–277.

Jamieson, G. A. and Vicente, K. J. (1998) "Modeling techniques to support abnormal situation management in the petrochemical processing industry," *Proceedings of the Canadian Society for Mechanical Engineering Symposium on Industrial Engineering and Management,* 3:249–256.

Jamieson, G. A. and Vicente, K. J. (2001) "Ecological interface design for petrochemical applications: Supporting operator adaptation, continuous learning, and distributed, collaborative work," *Computers and Chemical Engineering,* 25:1055–1074.

Janzen, M. J. and Vicente, K. J. (1998) "Attention allocation within the abstraction hierarchy," *International Journal of Human–Computer Studies,* 48:521–545.

Kirwan, B. and Ainsworth, L. K. (Eds.) (1992) *A Guide to Task Analysis*, London: Taylor & Francis.

Kohn, L. T., Corrigan, J. M., and Donaldson, M. S. (Eds.) (2000) *To Err Is Human: Building a Safer Health System*, Washington, DC: National Academy Press.

Kuo, J. and Burns, C. M. (2000) "A work domain analysis for VPN management," *Proceedings of the 2000 IEEE International Conference on Systems, Man and Cybernetics*, 1972–1977.

Leplat, J. (1991) "Organization of activity in collective tasks," in Rasmussen, J., Brehmer, B., and Leplat, J. (Eds.) *Distributed Decision Making: Cognitive Models for Cooperative Work*, New York: John Wiley & Sons.

Lind, M. (1990) *Representing Goals and Functions of Complex Systems* (90-D-381), Institute of Automatic Control Systems, Technical University of Denmark.

Lind, M. (1991) "Representations and abstractions for interface design using multilevel flow modeling," *Human–Computer Interaction and Complex Systems*, 223–243.

Lind, M. (1994) "Modeling goals and functions of complex industrial plants," *Applied Artificial Intelligence*, 8:259–283.

Militello, L. G. and Hutton, R. J. G. (1998) "Applied cognitive task analysis: A practitioner's toolkit for understanding cognitive task demands," *Ergonomics*, 41:1618–1641.

Miller, C. A., and Vicente, K. J. (1998) "Abstraction decomposition space analysis for NOVA's E1 acetylene hydrogenation reactor" (CEL-98-09), Cognitive Engineering Laboratory, University of Toronto.

Miller, C., and Vicente, K. J. (2001) "Comparison of display requirements generated via hierarchical task and abstraction-decomposition space analysis techniques," *International Journal of Cognitive Ergonomics*, 5:335–356.

Moiser, K. L. and Chidester, T. R. (1991) "Situation assessment and situation awareness in a team setting," in Queinnec, Y., and Daniellou, F. (Eds.), *Designing for Everyone*, 798–800, London: Taylor & Francis.

Moorefield, L. (1995) "Air Force information applications in the 21st century," *New World Vistas: Air and Space Power for the 21st Century — Information Applications Volume*, USAF Advisory Board.

Moradi-Nadimian, R. (2003) "Using ecological interface design to enhance a highway in the sky display," unpublished thesis, University of Waterloo.

Moradi-Nadimian, R., Griffiths, S. A., and Burns, C. M. (2002) "Ecological interface design in aviation domains: work domain analysis and instrumentation availability on the Harvard aircraft," *Proceedings of the 46th Annual Meeting of the Human Factors and Ergonomics Society*, 116–120.

Muller, M. J., Kuhn, S., and Meskill, J. A. (Eds.) (1992) *PDC'92: Proceedings of The Participatory Design Conference*, Cambridge, MA: Computer Professionals for Social Responsibility.

Muller, M. J., Tudor, L. G., Wildman, D. M., White, E. A., Root, R. W., Dayton, T., Carr, R., Diekmann, B., and Dykstra-Erickson, E. (1995) "Bifocal tools for scenarios and representations in participatory activities with users," in Carroll, J. M. (Ed.), *Scenario-Based Design: Envisioning Work and Technology in System Development*, 135–163, New York: John Wiley & Sons.

Nardi, B. A. (Ed.) (1996) *Context and Consciousness: Activity Theory and Human–Computer Interaction*, Cambridge, MA: MIT Press.

National Council of Welfare (1996) Gambling in Canada, Ministry of Supply and Services Canada, Ottawa: Government of Canada.

National Gambling Impact Study Commission (1999) Final Report, Washington, DC.

Nielsen, J. (1993) *Usability Engineering*, Boston: AP Professional.

Nimmo, I. (1995) "Adequately address abnormal situation management," *Chemical Engineering Progress*, 91:36–45.

Norman, D. (1988) *The Design of Everyday Things*, New York: Basic Books.

Object Management Group (2003) *OMG Unified Modeling Language Specification v1.5*, Needham, MA: Object Management Group.

Pawlak, W. S. and Vicente, K. J. (1996) "Inducing effective operator control through ecological interface design," *International Journal of Human–Computer Studies*, 44:653–688.

Perrow, C. (1990) *Normal Accidents*, Princeton, NJ: Princeton University Press.

Problem Gambling Research Group (1999) Community Impact of Increased Gambling Availability on Adult Gamblers — A four-year followup, University of Windsor. Available at http://venus.uwindsor.ca/pgrg/fyear.html.

Rasmussen, J. (1983) "Skills, rules, knowledge, signals, signs, and symbols, and other distinctions in human performance models," *IEEE Transactions on Systems, Man and Cybernetics*, 13:257–266.

Rasmussen, J. (1985) "The role of hierarchical knowledge representation in decision making and system management," *IEEE Transactions on Systems, Man and Cybernetics*, 15:234–243.

Rasmussen, J. and Jensen, A. (1974) "Mental procedures in real-life tasks: A case study of electronic troubleshooting," *Ergonomics*, 17:293–307.

Rasmussen, J. and Vicente, K. J. (1989) "Coping with human errors through system design: Implications for ecological interface design," *International Journal of Man–Machine Studies*, 31:517–534.

Rasmussen, J., Pejtersen, A. M., and Goodstein, L. P. (1994) *Cognitive Systems Engineering*, New York: John Wiley & Sons.

Reising, D. V. C., and Sanderson, P. M. (2002a) "Ecological interface design for Pasteurizer II: A process description of semantic mapping," *Human Factors*, 44:222–247.

Reising, D. V. C. and Sanderson, P. M. (2002b) "Work domain analysis and sensors II: Pasteurizer II case study," *International Journal of Human–Computer Studies*, 56:597–637.

Sanderson, P. M., Anderson, J., and Watson, M. (2000) "Extending ecological interface design to auditory displays," *Proceedings of the 2000 Conference of the Computer–Human Interaction Special Interest Group (CHISIG) of the Ergonomics Society of Australia (OzCHI2000)*.

Searcey, D. (1999) "Poll fuels effort to seek state aid to treat gamblers," *Seattle Times*. Available at http://problemgambling.com/pollsseek.html.

Selcon, S. J. and Taylor, R. M. (1990) "Evaluation of the situation awareness rating technique (SART) as a tool for aircrew systems design," *Situation Awareness in Aerospace Operations* (AGARD-CP-478), Neuilly Sur Seine, France.

Sharp, T. D. (1996) "Ecological interface design for the neonatal intensive care unit," unpublished thesis, University of Cincinnati.

Sharp, T. D. and Helmicki, A. J. (1998) "The application of the ecological interface design approach to neonatal intensive care medicine," *Proceedings of the 42nd Annual Meeting of the Human Factors and Ergonomics Society*, 350–354.

Sowb, Y. A., Loeb, R. G., and Roth, E. M. (1998) "Cognitive modeling of intraoperative critical events," *Proceedings of the IEEE Meeting on Systems, Man, and Cybernetics*, 2532–2538.

Stallings, W. (1999) *SNMP, SNMPv2, SNMPv3, and RMON 1 and 2, 3rd Ed.*, Reading, MA: Addison-Wesley.

Stoner, H. A., Wiese, E. E., and Lee, J. D. (2003) "Applying ecological interface design to the driving domain: The results of an abstraction hierarchy analysis," *Proceedings of the 47th Annual Meeting of the Human Factors and Ergonomics Society*, 444–448.

Thompson, L. K., Hickson, J., and Burns, C. M. (2003) WDA for blood glucose management. *Proceedings of the 47th Annual Meeting of the Human Factors and Ergonomics Society*, 1516–1520.

Tufte, E. R. (1983) *The Visual Display of Quantitative Information*, Cheshire, CT: Graphics Press.

Tufte, E. R. (1990) *Envisioning Information*, Cheshire, CT: Graphics Press.

Tufte, E. R. (1997) *Visual Explanations*. Cheshire, CT: Graphics Press.

Ulrich, K. T. and Eppinger, S. D. (2000) *Product Design and Development, 2nd Ed.*, Toronto: McGraw-Hill.

Vicente, K. J. (1990) "A few implications of an ecological approach to human factors," *Human Factors Society Bulletin*, 33:1–4.

Vicente, K. J. (1998) "Human factors and global problems: A systems approach," *Systems Engineering*, 1:57–69.

Vicente, K. J. (1999) *Cognitive Work Analysis: Toward Safe, Productive, and Healthy Computer-Based Work*, Mahwah, NJ: Erlbaum and Associates.

Vicente, K. J. (2001) "Cognitive engineering research at Risø from 1962–1979," in Salas, E. (Ed.), *Advances in Human Performance and Cognitive Engineering Research, Volume 1*, 1–57, New York: Elsevier.

Vicente, K. J. (2002) "Ecological interface design: Progress and challenges," *Human Factors*, 44:62–78.

Vicente, K. J., Moray, N., Lee, J. D., Rasmussen, J., Jones, B. G., Brock, R., and Djemil, T. (1996) "Evaluation of a rankine cycle display for nuclear power plant monitoring and diagnosis," *Human Factors*, 38:506–521.

Vicente, K. J. and Rasmussen, J. (1990) "The ecology of human–machine systems II: Mediating 'direct perception' in complex work domains," *Ecological Psychology*, 2:207–249.

Vicente, K. and Rasmussen, J. (1992) "Ecological interface design: Theoretical foundations," *IEEE Transactions on Systems, Man and Cybernetics*, 22:1–18.

Vidulich, M. A. and Hughes, E. R. (1991) "Testing a subjective metric of situation awareness," *Proceedings of the 35th Annual Meeting of the Human Factors and Ergonomics Society*, 1307–1311.

Waag, W. L. and Houck, M. R. (1994) "Tools for assessing situation awareness in an operational fighter environment," *Aviation, Space, and Environmental Medicine*, 65:A13–A19.

Watson, M., Russell, W. J., and Sanderson, P. M. (1999) "Ecological interface design for anesthesia monitoring," *Proceedings of the Australian/New Zealand Conference on Computer–Human Interaction (OzCHI99)*, 78–84.

Watson, M., Russell, W. J., and Sanderson, P. M. (2000) "Anesthesia monitoring, alarm proliferation, and ecological interface design," *Australian Journal of Information Systems*, 7:109–114.

Webb, R. D. G., McLean, D. N., and Matthews, M. M. (1998) Halifax Class Upgrade 2005: Cognitive Task Analysis of the ORO Position, Outline Method, Guelph, Ontario: Humansystems.

Webb, R. D. G., Matthews, M. L., Greenley, M. P., and Burns, C. M. (1993) Survey of Evaluation Methods for Command and Control Systems (Contract #W7711-2-7178/01-XSE). Toronto: DCIEM.

Woods, D. D. (1997) *The Theory and Practice of Representation Design in the Computer Medium (Ver 4.2),* Cognitive Systems Engineering Laboratory, The Ohio State University.

Yamaguchi, Y., and Tanabe, F. (2000) "Creation of interface system for nuclear reactor operation — Practical implication of implementing EID concept on a large complex system," *Proceedings of the XIVth Triennial Congress of the International Ergonomics Association and 44th Annual Meeting of the Human Factors and Ergonomics Society,* 3:571–574.

Zhang, Y., Drews, F. A., Westenskow, D. R., Forest, S., Agutter, J., Bermudez, J. C., Blike, G., and Loeb, R. (2002) "Effects of integrated graphical displays on situation awareness in anaesthesiology," *Cognition, Technology & Work,* 4:82–90.

Zinser, K. and Frischenschlager, F. (1994) "Multimedia's push into power," *IEEE Spectrum,* July:44–48.

INDEX

Distinctiveness principle of interface
design, 280
Duez, P., 200
DURESS project, 10–11, 141–142, 255

E

Ecological design, definition, 1–2
Ecological Interface Design (EID); *see also*
Visual forms in interface design; Work
Domain Analysis (WDA); *individual
case studies*
and Cognitive Work Analysis, 256–261
and Contextual Inquiry and Design,
267–273
and GOMS, 273–274
overviews, 1–12, 249–250, 287,
289–290
and Participatory Design, 276–278
and Situation Awareness, 265–267
and Task Analysis, 261–265
and Usability Evaluation, 281–286
and Use Case scenario design, 274–276,
277
and User Interface Principles, 278–281
Ecological psychology, 4
Endsley, M. R., 265
Energy and mass flow models, 143–144,
154, 156, 160–161, 172
Enhancement of existing designs, 102–103,
120–132, 138–139
Entertaining visualization development,
247–248
Environmental constraints on human
behavior, 4, 5–7; *see also* Work Domain
Analysis (WDA)
Eppinger, S. D., 16
Error prevention and correction criterion for
Usability Evaluation, 283, 285
Ethylene processing, 159–169, 174, 175–176
Evaluations of designs
acetylene hydrogenation reactor, 168
casino gambling, 246–247
and interface lifecycle, 253
network management, 193
nuclear power simulation, 151
oxygenation monitoring system, 210
pasteurizer, 158–159
patient monitoring in operating room,
220–221, 224, 225

thermal power generation system, 146
Usability Evaluation, 281–286
Use Case scenario design, 275–276, 277
Existing designs, enhancement of, 102–103,
120–132, 138–139
Explicitness criterion for Usability
Evaluation, 283, 284

F

Flexibility and control criterion for Usability
Evaluation, 283, 285
Flow models; *see also* Abstract Function
causal vs. functional, 175–176
for Contextual Inquiry, 268–269
information,158, 179–180, 182–183,
187–188, 199
mass and energy, 143–144, 155, 156,
160–161, 172
value relationships in social systems,
241–242, 247
Fluid catalytic cracking unit, 169
Focus and emphasis principle of interface
design, 280
Focus principle for Contextual Inquiry, 267
Frigate displays, 106–118
Fuel balancing case study for aircraft,
128–132
Functional Decomposition, 16–17
Functional flow models, uses of, 175–176
Functional information profiles, 102–103,
155, 156
Functional Purpose
and Abstraction Hierarchy, 16, 17, 18–20
acetylene hydrogenation reactor, 159
aircraft display cases, 121–122, 130, 196,
197
and design application, 85, 87, 88
diabetes management, 226
importance of determining, 133–134
naval vessel cases, 108
network management, 181–182,
185–186, 194
pasteurizer, 152, 155
and work domain model checklist, 41,
42
Functional vs. causal Abstraction Hierarchy,
27–28

vs. human factors methods, 19–20
large model management, 32–36
large refinery, 170–172
naval vessels, 106–120
network management, 181–184,
  193–194
nuclear power simulation, 149
origins of, 2
and other methods, 250–256
oxygenation monitoring system,
  204–205, 206, 207, 208, 209
pasteurizer, 153–155
patient monitoring in operating room,
  212–218
social constraint models, 31–32
summary checklist, 44–45
system of interest definition, 13–15,
  41–42
vs. Task Analysis, 263, 264, 265
telecommunications systems, 199

thermal power generation system,
  143–144
time constraints on, 106
transportation system challenges,
  133–138
Work Domain Model
  application to design, 85–104
  hierarchy combination in, 29–31
  ideal number of levels, 135–137
  large model management, 32–36
  social constraints on, 31–32, 33
  testing procedures, 36–41
Work Domains level in Cognitive Work
  Analysis, 256
Workspace visual form level, 48

Z

Zhang, Y., 221–224